THE PO
CALCUL

Lionel Carter is a qualified Chartered Mechanical Engineer. He is Principal Lecturer in Management Science at Slough College of Higher Education and a visiting lecturer at Brunel University. Previously he was an operational research consultant in industry. He has acted as an adviser to the EITB in developing an investment appraisal training programme.

Dr Eva Huzan is Head of the Computing Division at Slough College of Higher Education. Previously she worked as a physicist and computing lecturer in industry. She has carried out research in computing and physics at the London School of Economics and Political Science, and Queen Mary College, University of London.

The authors have also collaborated in writing Teach Yourself books on *Computer Programming in BASIC*, *Microelectronics and Microcomputers* and *Computer Programming with the Commodore 64*.

TEACH YOURSELF BOOKS

THE POCKET CALCULATOR

L. R. Carter and E. Huzan

TEACH YOURSELF BOOKS

Hodder and Stoughton

First published 1979
Reissued (with additions) 1983
Fifth impression 1984

British Library Cataloguing in Publication Data

Carter, L. R.
The pocket calculator.—(Teach yourself books)
1. Calculating machines
I. Title   II. Huzan, E.
510′.28   QA75

ISBN 0–340–33836–9

Printed and bound in Great Britain for
Hodder and Stoughton Educational,
a division of Hodder and Stoughton Ltd,
Mill Road, Dunton Green, Sevenoaks, Kent,
by Richard Clay (The Chaucer Press) Ltd, Bungay, Suffolk

Contents

Acknowledgments

We are indebted to the following companies for supplying us with information on the operation of their calculators: C B M Business Machines, Casio, Dixons Photographic, Rockwell International, Sinclair Radionics, Texas Instruments.

We also thank Hewlett-Packard for allowing us to base equations 7.4, 7.5, 7.6 on material from one of their applications manuals.

We are indebted to David Croft and Macdonald and Evans (Publications) Ltd. for permission to reproduce the statistical table (Appendix D) from *Applied Statistics for Management Studies*.

Introduction

Calculators are being used increasingly widely in many different fields. Students may now use them in various examinations and at school, college and in industry. Scientists and engineers find them more suitable for many calculations than the traditional slide rule while they are invaluable to accountants and managers when analysing a company's operations.

The aim of this book is to help the reader become familiar with the techniques of using a pocket calculator. However, you should use this book in conjunction with the manufacturer's booklet which is supplied with your calculator in order to ensure that you are using the functions correctly and using the necessary keying sequence for the type of logic built into your calculator. The manufacturer's booklet will also give you the operating accuracy limitations and calculation range.

The book is divided into two parts. Part I, chapters 1 to 6, explains the principles involved in using a calculator, and gives sufficient background to enable you to work out the problems set in Part II, chapters 7 to 10. The first three chapters are complete in themselves, in that chapter 1 discusses choosing a calculator, chapter 2 demonstrates the use of the calculator for simple calculations using the basic functions, and chapter 3 explains the sources of error and methods of correcting mistakes in entering wrong data or functions.

Chapter 4 gives the mathematical background to many of the scientific functions. Chapters 5 and 6 cover the use of the calculator for solving problems involving trigonometry, volumes and area.

Chapter 7 will be particularly useful for those interested in financial and management problems but it also deals with statistical applications and some of these will be of interest to the scientist and engineer. Chapter 8 explains the use of the calculator in solving financial and accountancy problems. Chapter 9 is concerned with the calculations associated with management problems. Problems in science and engineering are categorised and discussed in chapter 10. The emphasis in this chapter is on the types of solution rather than on the calculator functions to be used. Thus drawing-up tables and using graphs is discussed with reference to the calculator, as well as the evaluation of formulae for different types of problem which requires the interpretation of experimental results.

Keying tables are used throughout the book to show the sequence of keys to be depressed to solve a particular problem, together with intermediate and final results, and comments where appropriate. In some cases, even though the arithmetic is simple, the complete keying sequence has been given; in other cases some mental arithmetic has been carried out before or during the calculation to show how the amount of data to be keyed in can be reduced. In Part I some of the keying tables are included in the body of the text as examples of the methods that can be used. You can practise these methods by using the calculator for the problems in Part II where the keying tables are presented separately in Appendix A. Solutions to additional problems in the text are given in Appendix B.

Programmable calculators

Programmable calculators are useful whenever the user wants to perform the same calculation over and over again with different values. A programmable calculator allows the user to enter the required keying sequence into the calculator's memory. Having entered the keying sequence in 'program mode' the user presses a RUN/STOP key. This causes the calculator to follow the sequence of key depressions stored in the memory, only stopping when a value is required to be entered from the keyboard. On entering the value the user presses the RUN/STOP key again and the calculation continues automatically.

There are two main advantages in using a programmable calculator for repetitive calculations. Firstly, the keying sequence has only to be entered once. Secondly, the speed of arriving at the answer is considerably increased, as the calculator does not have to wait for the user to press the 'next' key.

The manner of switching to 'program mode' and running a program varies from calculator to calculator. However, the sequence of program instructions would be very similar to those used with a non-programmable calculator. Thus the explanations of the function keys given in this book also apply to programmable calculators and the keying sequences described in this book would be substantially the same for a programmable calculator.

A simple requirement to convert feet to metres by multiplying by the factor 0.3048 might be programmed as follows:

> switch to 'program mode'
> key in: ×
>   0.3048
>   R/S (the RUN/STOP key)
> switch out of 'program mode'

To run the above program the number of feet are keyed in and

then the R/S key pressed. The number of feet will then be automatically multiplied by 0.3048 and the calculator will stop with
the result displayed. To convert a different number of feet the
new value is keyed in and the R/S key pressed again. The calculator will convert this value and stop once again with the metric
equivalent in its display.

The number of keying instructions that can be stored in a
programmable calculator's memory depends upon the model.
In practice, a programmable calculator is especially useful when
the keying sequence is long. Even though the calculator is programmable it can become very tedious to have to key in a long
program. For this reason, a means of 'saving' a program is often
provided. One simple method is for the calculator to have a
'continuous memory'; that is, the program currently in the
memory is not lost when the calculator is switched off. The program is immediately available for use when the calculator is
turned on again.

Continuous memory, however, only allows the most recently
entered program to be retained (or the last few, depending on the
size of memory). For longer term storage the more elaborate
programmable calculators allow the program to be 'saved' (e.g.
transferred) on to a magnetic card or a cassette tape. In some
instances, sets of programs are also available in solid state
modules that plug into the calculator.

The instruction books issued with programmable calculators
will tend to assume a familiarity with the function and use of the
keys normally found on scientific pocket calculators. The instruction books will mainly concentrate on the principles of
using the programming features. A knowledge of the function
keys and keying techniques as discussed in this book is still
therefore necessary, whether the calculator is being used in
direct or programmable mode.

Part I
Background and Principles

1 Choosing a calculator

1 Introduction

The potential purchaser of a pocket calculator is faced with a wide range of different specifications offering numerous facilities. This chapter describes the basic features of calculators and is intended to help you choose your model. The areas to be discussed have been classified into three broad aspects, the technical (keyboard, display, etc.), power (battery, mains, etc.) and physical (size, weight, etc.).

2 Technical aspects

The newcomer to pocket calculators is faced with a certain amount of jargon that has evolved along with the product. Apart from knowing the meaning of the terms, it is important, if a suitable choice is to be made, to understand their significance. It is obviously not necessary to understand the inner workings of the calculator and therefore the approach here is to describe only those aspects that affect the use of the finished product. Leaving aside the question of power, which is dealt with later, you can consider the calculator as comprising two units, the keyboard and the display.

2.1 *Keyboard*

The basic keys that must be on all calculators are the ten numbered 0 to 9, a decimal point key and four others for the arithmetic functions of addition, subtraction, multiplication and division. The use of the keys varies according to the logic built into the calculator, as discussed in the following section.

2.1.1 *Different types of logic*

Very simple calculators are likely to have either algebraic or arithmetic logic while advanced 'scientific' calculators are likely to have algebraic logic or Reverse Polish Notation (RPN). As the majority of calculators have algebraic logic, this book is based upon that type of calculator. However, this section briefly discusses the difference between the three types.

With algebraic logic machines numbers and operators ($+$, $-$, $\times$, $\div$) are entered in the same sequence as they are written and the calculation is terminated by using the equals key,

i.e. $\qquad\qquad\qquad 4 + 6 - 2 =$

The keying sequence with this type of logic is the most natural and requires no adaptation on the part of the user.

A variation encountered on some calculators is that $\times$ and $\div$ operations in a chain calculation are performed first (see p. 17).

Arithmetic logic machines can usually be recognised by the presence of the equals symbol on both the add and subtract keys (i.e. $\pm$, $\underline{-}$). With these calculators multiplication and division are done in the same natural way as for algebraic logic, but for addition and subtraction the number is entered first followed by the operator ($+$ or $-$). For example, to calculate $10 - 4$ you need to enter $10 \pm 4 \underline{-}$.

Many of the earlier calculators were of this type but now arithmetic logic is largely confined to desk calculators.

Reverse Polish Notation requires the number to be entered first, followed by the operator for all functions. The keying sequence for,

$$(3 + 5) \times 4 - 6$$

would be

$$3 + 5 + 4 \times 6 -$$

At first this method seems odd and difficult, although technically it does have some advantages leading to fewer key depressions, compared to other calculators. It is extensively used on advanced scientific calculators where the complexity of the calculations justifies mastering the notation.

To sum up, arithmetic logic calculators offer no advantages and should be avoided. Reverse Polish calculators offer tangible advantages in advanced and complex calculations which justify their use by scientists and engineers. For the ordinary user and the student an algebraic logic calculator is the natural choice.

2.1.2 *Keys*

This section outlines the meaning of the most common keys to be found on calculators. More detailed descriptions of their use are given in chapters 2, 4 and 5. It is assumed that the reader is familiar with the meaning of the numeric keys (0 to 9), the basic operators ($+$, $-$, $\times$, $\div$) and the equals ($=$) keys.

The decimal point key is depressed at the appropriate time when entering a number. Final zeros after the decimal do not need to be entered, e.g.

<pre>
        for 120.17 enter 120.17
        for 100.0 enter 100
        for £10.57 enter 10.57
        for   4p enter   .04 (in £s).
</pre>

It is unnecessary here to enter a zero prior to the decimal point. The entry of final zeros after the decimal is not wrong, just unnecessary.

C, CE, CM

A clear key or keys are provided for:

(a) clearing the calculator of all previous entries and results, usually marked C,
(b) clearing the last entry, marked CE,
(c) clearing the contents of the memory only, marked CM or MC.

Sometimes one clear key fulfils requirements (a) and (b), i.e. the last entry is cleared by depressing the key once and the display is cleared by depressing the key twice.

M, STO

A memory allows a value to be stored and recalled for later use in a calculation. It may therefore save writing down intermediate answers. The contents of the memory are lost when the calculator is switched off (except in the case of a few advanced scientific calculators). Most calculators have a single memory or store enabling only one value to be recalled. Some calculators, however, have more than one memory.

The memory is used in one of two ways. One type of calculator allows the value stored in the memory to be altered directly by means of M+ (add to memory) and M— (subtract from memory) keys. The other type of calculator simply allows a number to be stored (usually by means of a STO key); to alter the value of this number, it has to be recalled, manipulated and returned to store.

The key to recall the memory is usually marked MR or RCL. A further key provided on some calculators allows the value of the current calculation and the contents of the memory to be exchanged; this key is designated MEX or M ↔ ×.

A few calculators indicate on the display when the memory is in use.

( )

The use of the memory is often required because the expression being evaluated contains brackets. Rather than providing several memories an alternative is to include parentheses (brackets) facilities. There are two keys, one 'to open brackets' marked (, and one 'to close brackets' marked).

More than one level of parentheses may be provided. If only a single level of parentheses is available, it should be appreciated that it can be used as many times as required in one calculation, provided that any one set of brackets is closed before another is opened.

K

Many calculators have a constant facility which allows a number that is to be repeatedly used in a current calculation to be entered only once. Sometimes there is a specific key for this but more usually it is built into the logic. When it is built into the logic, there is no specific key involved but the manner of operating the keys provides the constant facility.

It is important to appreciate that a constant facility can only be used on successive key entries. It is not a memory and cannot be referred back to at a later stage. However, a memory can be set up to hold a 'constant' for use whenever required.

%

Some basic calculators are provided with a percentage key (%). The need for % calculations arises in many circumstances, particularly applications in the retail trade where such a key may be useful to someone unfamiliar with decimals. Scientific calculators do not usually have % keys, however, as it is judged that their users will be familiar with the idea of entering the decimal equivalent of a %.

The depression of the % key usually causes the calculator to treat the previously entered number as a %, and displays the result directly without the necessity to depress the equals key.

F

In order to keep the keyboard compact, many calculators attach two uses to some of the keys. Prior to using the key for its alternative purpose, the F (function key) has to be pressed. This is analogous to the shift key on a typewriter, where each key serves two purposes. A compact layout is therefore obtained at the expense of having to use two keystrokes for some of the functions.

$\frac{1}{x}$, x^2, $\sqrt{x}$

These keys (reciprocal, square and square root of number) are the most common keys to be added to a basic calculator to make it equivalent to the traditional slide rule. Calculators with these facilities, therefore, are a compromise between the simple calculator and the more specialised scientific models.

2.1.3   *Specialised keyboards*

As facilities are added to the basic calculator, manufacturers have the option of providing differing keyboards to meet specialised needs. Two types of keyboard have evolved, the general 'scientific' and the financial.

The financial keyboard allows annuities, present values, etc., to be evaluated directly and meets the need of the financial specialist. However, these options are very specialised indeed, and by including them some general functions may need to be omitted.

The 'scientific' calculator is more common and is the more flexible in its capabilities. Unlike the financial calculator, it is not restricted to one area of application. This book is based on a general-purpose scientific calculator, with the facilities listed in the next section.

A variation on the 'scientific' calculator that is growing in popularity is one having statistical functions (e.g. standard deviation, automatic summation of entries). If you are thinking of buying this type of calculator you need to check on the continued availability of the memory. Some calculators make automatic use of the memory in computing the statistical functions, thereby depriving the user of the memory.

2.1.4   *Facilities covered by this book*

In the examples and problems, it is assumed the following functions are always available; their meaning and use are explained in chapters 2, 4 and 5.

$$+, -, \times, \div$$
$$\text{M}+, \text{M}-, \text{MR}, \text{MEX or M} \leftrightarrow \times \quad \Big\} \text{chapter 2}$$

$$\left. \begin{array}{l} \dfrac{1}{x} \\[2mm] x^2, \sqrt{x} \\[1mm] ln, e^x \\[1mm] \sin, \cos, \tan \\[1mm] \sin^{-1}, \cos^{-1}, \tan^{-1} \text{ (arcsin, arcos, arctan)} \\[1mm] \pi \end{array} \right\} \text{chapters 4 and 5}$$

Mention is made of the following functions, with an explanation where necessary, but it is not assumed that these functions are available in further examples and problems.

$$\%$$
$$\log, 10^x$$
$$y^x, \sqrt[x]{y}$$
$$+/- \text{ (change sign)}$$
two memories
$$(\ )$$

2.2 Display

2.2.1 Type of display

There are three types of display in common use. One of these uses light-emitting diodes (LED), which show up as small red figures, and often has magnification lenses fitted in front of the digits. A common alternative is the green fluorescent display which is larger and has the advantage of consuming less power. The third type of display is the liquid crystal (LCD) and this type also uses very little power. Liquid crystal displays allow very thin calculators to be built. A possible disadvantage of the LCD display is their relatively slow reaction time, but this is usually unimportant for routine calculations.

2.2.2 Contents of display

Calculators vary in the number of digits it is possible to display, from five to eight being typical. Most calculators display the answer in floating point notation, that is, the decimal point moves automatically across the display to take up its correct position.

An alternative form of display, scientific notation, is available additionally or alternatively on some calculators. In scientific notation any number is split into two components, a mantissa and an exponent. The mantissa when multiplied by 10 to the power of the exponent expresses the required number, e.g.

$$127.2 = 1.272 \times 10^2$$

(here 1.272 is the mantissa and 2 is the exponent; the latter also is referred to as the index or power)

$$130000 = 1.3 \times 10^5$$
$$0.0125 = 1.25 \times 10^{-2}$$

It is common for only the mantissa and exponent to be displayed, the presence of the number 10 being implied. For the above examples, the displays would be:

1.272	02
1.3	05
1.25	−02

Calculators using this notation are usually designed to allow a two-digit exponent (−99 to 99) to be used.

The implications of the number of digits in the display are discussed further in chapter 3.

The display on some calculators is used also to show when the memory is in use, when the battery is failing and when an unacceptable computation is attempted. The latter may be indicated by a letter E to the left of the display.

3 Powering a calculator

The principal means of powering a pocket calculator is with batteries but some calculators can be powered also from the mains via an adaptor.

3.1 *Disposable batteries*

There is a range of batteries available for calculators. It is advisable to choose leakproof batteries and, if the calculator is to be used for several hours at a time, it may be economic to buy the higher priced alkaline-type batteries. An alternative available in some sizes is the special 'calculator' battery. This is likely to be dearer than the standard battery but cheaper than the alkaline.

3.2 *Rechargeable batteries*

To reduce the running cost it is possible to use rechargeable batteries. Some calculators come supplied with rechargeable batteries and a charging unit, with others they are extra.

Even if there is no mention of rechargeable batteries in your instruction booklet it is possible to use rechargeable batteries. The drawback is that, unless there is a recharging socket on the calculator, they cannot be recharged *in situ* but will need to be removed and charged in a separate unit. In general, therefore, it is better to take this aspect into account before buying a calculator.

A drawback with rechargeable calculators is that they cannot be replenished instantaneously but need to be charged for several hours, for example overnight.

3.3 *Mains adaptors*

The mains adaptor may be used solely for running the calculator off the mains or it may be used also as a battery charger.

If portability is a major consideration, it should be realised that adaptor and chargers are heavier than calculators and, with the mains plug fitted, more bulky. Thus, if you want to use your calculators in hotels, etc., for several weeks, remember that the space taken up in a brief case is more than twice the size of the 'pocket' calculator.

4 Physical aspects

4.1 *Dimensions*

The physical dimensions of pocket calculators vary a great deal. This is obviously a personal choice and depends on such factors as:

  (*a*) the main location and method of usage (home, office, college, in the field), (*b*) method of transportation (pocket, brief case, room to room),

(*c*) size of hand, fingers, dexterity.

The design features that have the major influence on calculator size are the type of display, the capacity of the battery and the keyboard layout. The smallest calculators would have LCD, with the keys performing two functions where necessary.

Some small calculators tend to slip around if used on a flat surface unless they have non-slip pads.

4.2 Keyboard

4.2.1 Keyboard touch

The way keys have to be used may influence the choice of calculator. The necessary action may be:

(*a*) a firm positive depression of the keys, which leads to a 'click',

(*b*) a light depression of the key,

(*c*) a light touch of the key (not favoured if keys are close together, as is the case with the smaller calculator).

If keys require a light depression or touch it may sometimes be difficult to know whether the entry has actually been made. Also, a light touch may lead to the key being depressed twice accidentally. Some people, therefore, prefer the positive action. This is very much a personal choice.

4.2.2 Layout

The most appropriate layout of the keys depends on how the calculator will be used. If the full range of facilities is to be used regularly, then it is preferable to have as many separate keys as possible, so as to avoid having to use double-function keys. This will, however, increase the size of the keyboard and the cost. A smaller and cheaper calculator which, by having double-function keys, offers the same facilities may be preferable. If only some of the functions will in general be used, then their relative position for ease of entry may determine the choice of calculator. This is best tested by doing some trial calculations.

Some calculators have the keyboard at the side of the display. These can be awkward to use in the hand, particularly for left-handed people.

4.2.3 *Computing speed*

The speed of the calculator is determined by its circuitry rather than the keyboard.

Most calculators are 'instantaneous' for the simple arithmetic functions but have a slight and perceptible delay for the scientific functions. Therefore, it is advisable to try out a calculator before making a final decision to buy it.

5 Purchaser's check list

(a) Type of logic (arithmetic, algebraic, RPN).
(b) Type of display.
(c) Number of digits.
(d) Type of notation used.
(e) Additional indicators in display.
(f) Means of power.
(g) Charger/adaptor extra.
(h) Memory (M+, M−, MR) or Store (STO, RCL).
(i) Memory used by other functions.
(j) Parentheses (number of levels).
(k) Single or dual function keys.
(l) Automatic constant.
(m) Keys, as required.

2 Use of common facilities

1 Introduction

The precise sequence of keying varies with the model of calculator but the instruction booklet supplied should make the method of performing the four basic functions (+, −, ×, ÷) clear. This chapter develops the use of basic facilities; the use of keys associated with 'scientific' functions is discussed in chapters 4 and 5. The keying sequence shown in the tables is for algebraic logic with no precedence of the × and ÷ operators (this is a common form of logic).

The display is shown on the keying tables only where it will be helpful to the reader, i.e. where intermediate results are displayed and when the memory is recalled. The display after a number has been keyed in has been left blank (as it is obviously that number) except where, for instance, the use of the exponential notation is illustrated.

The number in the display column has been truncated (see chapter 3) to five significant digits in most cases; more digits may be shown if there is a significant decimal place or greater accuracy needs to be noted for, say, later keying in. The answers have been rounded (see chapter 3) where appropriate.

2 Arithmetic function keys

The arithmetic function keys are marked +, −, ÷, ×. The following are examples of 'one stage' calculations:

(a) 5 × 7   (b) 6 ÷ 3   (c) 10 ÷ 6
(d) 9 − 10   (e) 3 + 12.2   (f) −5 − 3

Calculations (a) and (b) are straightforward. In calculation (*c*) look at the final decimal place. Because the answer is 1.6666 with the 6 recurring, a mathematician would, in practice, 'round up' the final 6 to 7. Usually calculators do not 'round up'; if the answer is too long for the display it is truncated (chopped off). It may be important if you are buying quantities of wallpaper, wool, wood, etc., to round up the answer rather than truncate it.

Calculation (*d*) should give an answer of -1.0, check that the minus is indicated and ensure that the method of displaying a negative quantity is known.

Calculation (*e*) introduces an answer having one decimal place. Note whether all the digits are displayed across the display, in this case several zeros, or whether only the significant answer of 15.2 is shown.

Calculation (*f*) raises the problem of initially entering -5. It is important to be familiar with the method of entering an initial negative quantity.

2.1 *Chain arithmetic*

When you are familiar with the above types of calculation, chain calculations can be considered. For straightforward chain calculations such as,

$$5 \times 4 \times 3 \times 6 \times 2$$

there is no problem. Problems arise when there is a mixture of arithmetic functions in the calculation. For example, $5 \times 4 - 2$ is ambiguous, it could mean $(5 \times 4) - 2$ or $5 \times (4 - 2)$, namely 18 or 10, depending upon the sequence of steps in the calculation.

It may be necessary to enter the numbers in a different sequence to the way they are written. For example, the second interpretation above, leading to an answer of 10, would be entered *as though* it were written $4 - 2 \times 5$. The depression of the $\times$ key signifies the end of the second number, causes the calculator to act on the subtraction and presets the multiplication.

Some calculators give precedence to $\times$ and $\div$ operations over $+$ and $-$, which result in a different answer for chain calculations unless the equals key is depressed prior to a × or − operation, i.e.

with no precedence −1 + 5 × 2 = 8 (since (−1 + 5) × 2
is assumed

with × precedence −1 + 5 × 2 = 9 (since −1 + (5 × 2)
is assumed.

This book assumes no precedence is built into the calculator. If a calculator having such precedence is being used, then greater efficiency in the order of operations may be possible for some of the examples illustrated.

2.2 *Example 1*

What is the total cost of five panels, 2 m × 3 m of blockboard and plastic laminate, when blockboard cut to size costs £2.00 per sq. metre (m²) and laminate £1.50 per sq. metre?

This calculation can be done conventionally in stages, if required, to give:

Cost of blockboard = 2 × 3 × 5 × 2 = 60
Note result (£60.00) and then calculate,

Cost of laminate = 2 × 3 × 5 × 1.5 = 45
Add previous 60 to current display (45) to obtain a total cost of £105.

However, arithmetically, the calculation can be written:

Area per panel × number of panels × (cost of blockboard per m² + cost of laminate per m²), i.e.

$$3 \times 2 \times 5(2 + 1.5)$$

Assuming the calculator does not have a memory, an approach would be to write down the answer to 3 × 2 × 5, then add 2 and 1.5 to get 3.5. This result could then be multiplied by the previously noted value by entering × 30 to obtain the final answer of 105. However, the calculation can be done, without a memory and without noting down intermediate answers, if the brackets are evaluated first, as shown in Table 2.1.

Table 2.1 Example 1

Keying	Display	Comments
2		
+		
1.5		
×	3.5000	result after +
5		
×	17.5	result after first ×
2		
×	35.	result after second ×
3		
=	105.	final result

2.3 *Division*

Calculations involving division can lead to different approaches to evaluation. Provided all the numbers in the divisor (bottom line) are being multiplied together, the top line (numerator) can be worked out first and then the answer divided by each of the values in the divisor, e.g.

$$\frac{2 + 7 - 3}{4 \times 5}$$

The numerator can be evaluated to give 6. This result is then divided by 4 and then divided again by 5 to give finally the answer 0.3.

In the case where the divisor *does not* consist only of multiplications, you are likely to need either a memory facility or to write down intermediate results. Occasionally, it may be possible to turn the expression 'upside down' and evaluate, finally turning the answer 'right way up' by using a reciprocal key.

i.e. consider
$$\frac{6 \times 8}{3 + 4 + 5}$$

Turning this expression 'upside down' gives

$$\frac{3+4+5}{6 \times 8}$$

This can be evaluated now as before without writing down intermediate results. The sequence being $3 + 4 + 5 \div 6 \div 8$, the result (0.25) however is 'upside down'. The correct answer (4) can be found by taking the reciprocal $\left(\frac{1}{x}\right)$ of the result.

Arithmetically,

$$\frac{6 \times 8}{3 + 4 + 5} = \frac{1}{\dfrac{(3 + 4 + 5)}{6 \times 8}} = \frac{1}{.25} = 4$$

The reciprocal key $\left(\frac{1}{x}\right)$ is discussed in section 6.

2.4 Example 2

The following can be evaluated using straightforward chain arithmetic, by first evaluating the brackets.

$$\frac{5.76(9.2 + 4.3)}{6.2 \times 7.3}$$

The keying sequence is given in Table 2.2 and gives the answer 1.7181.

Table 2.2 Example 2

Keying	Display	Comments
9.2		
+		
4.3		
×	13.5	result after +
5.76		
÷	77.76	result after ×
6.2		
÷	12.541	result after first ÷
7.3		
=	1.71807	final result 1.7181

2.5 *Example 3*

In a shop, the time taken to deal with customers' enquiries on the last seven occasions was 15, 12, 10, 8, 19, 11, 13 minutes.
What was the average time taken?

Table 2.3 Example 3

Keying	Display	Comments
15		
+	15.	
12		
+	27.	
10		
+	37.	
8		
+	45.	
19		
+	64.	
11		
+	75.	
13		
÷	88.	cumulative time
7		
=	12.571	minutes, i.e. 12 minutes
−		
12		
×	0.57142	decimal part of minutes
60		
=	34.285	seconds, i.e. 34 seconds

The keying sequence is shown in Table 2.3 and gives the answer 12.57 minutes, which has been converted in the table to 12 minutes 34 seconds by multiplying the decimal part by 60.

3 Sign exchange key

Some calculators have a sign exchange key, usually marked +/−. This can be useful in so far as it often allows you to think of the problem in the 'normal' way, whereas the algebra requires a different approach. For example, if you were working out the current balance in your bank account, you would normally add your total expenditure, obtained from your cheque stubs, and *then* deduct them from the initial balance. Arithmetically if the initial balance is +£100, sundry expenditure might lead to this calculation,

$$100 - 5 - 15 - 10 - 12, \text{ i.e. } 58$$

However, the logical way to approach this problem is to add the expenditure first from your cheque book. This gives the following sequence,

$$-(5 + 15 + 10 + 12) + 100$$

In this case, the sign of the total expenditure needs to be changed i.e. +42 to −42. A 100 is then added to obtain the answer 58. A change sign key will allow this to be done; the keying sequence is shown in Table 2.4.

Table 2.4   Use of sign exchange key

Keying	Display	Comments
5		
+	5.	
15		
+	20.	
10		
+	30.	
12		
=	42.	total expenditure
+/−	−42.	
+		
100		
=	58.	

4   Clear keys

The habit of clearing the whole calculator before starting on a new calculation is important.

If there is a memory facility, it is important to ensure that this has been cleared as well before continuing on to different problems.

The clear last entry key is a useful facility as it allows you to cancel the last number entered and to continue with alternative data without going back to the beginning. This key is most widely used to overcome keying mistakes. Further discussion on overcoming incorrect entries is given in chapter 3.

5 The memory facility

The use of memory keys is straightforward but the following points need to be considered. If an M— key is not available, the same effect can be achieved by changing the sign of the display, as previously explained, using the +/— key and then adding it to the memory using M+.

It is necessary to establish the action of the MR key. Some calculators automatically clear the memory when it is recalled. In this case to retain the value in the memory, the M+ needs to be used to reintroduce the displayed results back into the memory.

Usually, if the key is marked M+ the implications are that the current display is added to the existing contents of the memory. Some calculators have store keys (STO) which simply retain the current display in place of the previous contents of the store. To add to the store in these circumstances requires that the current display has added to it the recalled value of the store, and the revised total is then stored.

5.1 *Memory exchange*

The MEX key is useful in some calculations as this facility enables the memory to be accessed for temporary reference purposes. Depressing the memory exchange key again brings the original values back into the display and memory respectively, allowing the calculation to continue.

If there is no memory exchange key, it is possible by manipulation to exchange the display with the memory. The routine is given in Table 2.5, where the value in the display is represented by D and the value in the memory by M. At the end of the manipulation the display contains the negative value of the original memory. Depending on other available keys, you would then need to change the sign as previously explained.

Table 2.5   Memory exchange manipulation

Keying	Display	Comments
	D	M in memory
−MR = M+	D − M	M + (D − M) = D in memory
−MR =	−M	D still in memory

5.2   *Example 4*

The following example requires the use of a memory facility.

$$\frac{(5.76 + 9.6)}{(8.2 + 7.25)} \times 5.04 \times 3.6$$

The keying sequence is given in Table 2.6 and gives the result 18.038.

Table 2.6   Example 4

Keying	Display	Comments
8.2		
+		
7.25		
=	15.45	
M+		retain for later use
5.76		
+		
9.6		
×	15.36	(5.76 + 9.6)
5.04		
×	77.414	
3.6		
÷	278.69	
MR	15.45	
=	18.038	

Table 2.7 Requirement for memory or parentheses in solving selected problems

Problem	Facilities Required
6.2(5.23 + 7.42)	no memory, no parentheses
(1.7 + 6.2)(5.23 + 7.42)	one memory or one level of parentheses
(1.7 + (6.2 × 4.1))(5.23 + 7.42)	one memory or one level of parentheses
$\left(\dfrac{1.7}{2.3} + (6.2 \times 4.1)\right)(5.23 + 7.42)$	one memory (used twice) or one level of parentheses (used twice)

5.3 *Parentheses*

Table 2.7 shows the memory or parentheses requirements for a few sample problems. These illustrate how even a single memory or one level of parentheses is an important feature on a calculator.

5.4 *Example 5*

The calculation illustrated in Example 4 can be carried out with a single level of parentheses in place of a memory.

The keying sequence is given in Table 2.8.

Table 2.8 Use of parentheses

Keying	Display	Comments
5.76		
+		
9.6		
×	15.36	
5.04		
×	77.414	
3.6		
÷	278.69	
(		division will apply to entire bracket
8.2		
+		
7.25		
)	15.45	evaluation within brackets
=	18.038	

6 Reciprocal key

The reciprocal key can be an advantage in chain calculations where, by rearranging the sequence of keying, intermediate answers need not be noted. This aspect, which is the more useful feature of a reciprocal key, has been covered under arithmetic functions.

6.1 *Example 5*

$$\frac{7.2 \times 3.6}{4.2 + 1.7}$$

can be regarded as:

$$\frac{1}{4.2 + 1.7} \times 7.2 \times 3.6$$

which, with a reciprocal key, becomes a chain calculation, as shown in Table 2.9 giving the answer 4.3932.

Table 2.9 Use of reciprocal key

Keying	Display	Comments
4.2		
+		
1.7		
=	5.9	
1/x	0.16949	reciprocal of 5.9
×		
7.2		
×	1.2203	
3.6		
=	4.3932	

7   Percentage key

The use of this key is illustrated in the following Examples.

7.1   *Example 7*

What is the VAT payable on a bill of £23.63? (VAT rate = 8%)
   The keying sequence for a calculator having a % key is given in Table 2.10 and gives the answer £1.89. Without a % key, the 8% is treated as 8 hundredths or 8 per hundred. If you are unsure

Table 2.10   Using the % key

Keying	Display	Comments
23.63		bill
×		
8		VAT rate
%	1.8904	treat as %; Ans. = £1.89

of the decimal point, this can be entered using three keying operations, i.e. 8, ÷, 100. Alternatively, the value .08 can be entered directly. These two alternatives are shown in Tables 2.11 and 2.12, respectively. As can be seen, using the decimal form directly leads to fewer key depressions.

Table 2.11   % calculation without the % key

Keying	Display	Comments
23.63		bill
×		
8		VAT rate
÷	189.04	
100		i.e. 8% = 8 ÷ 100
=	1.8904	£1.89

Table 2.12   Entering % direct as a decimal

Keying	Display	Comments
23.63		bill
×		
.08		VAT as a decimal
=	1.8904	£1.89

7.2   % *conversion rules*

The rule to follow when converting a % to a decimal is:

Drop the % sign and move decimal point two places to left.

Conversely, when converting a decimal to a %:

Move decimal two places to the right and add % sign.

i.e. $12.5\% = .125$

and $.762\ \ \ = 76.2\%$

7.3   *Example 8*

J. Smith has a 6.8% increase on his monthly salary of £376.72. What is his new monthly salary?

The increase in the salary can be calculated by multiplying by .068. This then needs to be added to the original salary, i.e. original salary × 1. The revised salary can therefore be calculated directly by multiplying by 1.068.

Table 2.13 illustrates the method.

Table 2.13   Calculating % wage increase using decimal keying

Keying	Display	Comments
376.72		salary
×		
1.068		(1 + 0.068) done mentally
=	402.336	revised salary = £402.34

7.4 % *increase, decrease rules*

To increase a given quantity by a %, multiply by (1 + decimal equivalent). To reduce a quantity by a %, multiply by (1 − decimal equivalent).

The remainder of the book treats percentages directly as decimals. If many % calculations are to be undertaken in practice, it will be worthwhile to become familiar with this approach, and the reader is referred to *Teach Yourself Arithmetic* (L. C. Pascoe).

8 Conversions

To convert from one unit of measurement to another should not present a problem. It is a matter of multiplying (or dividing) by an appropriate factor. Some specialised calculators have preprogrammed conversion keys but the majority do not. A set of useful conversion factors is given in Appendix C.

8.1 *Example 9*

It is required to convert £25, £40 and £127 into their dollar equivalents, where $1.75 = £1.00.

Table 2.14   Conversion of pounds to dollars

Keying	Display	Comments
1.75		conversion factor
×		
25		£25.00
=	43.75	$
40		£40.00
=	70.	$
127		£127.00
=	222.25	$

Table 2.14 shows the keying sequence, using the constant facility. The answers are $43.75, $70 and $222.25, respectively.

9   Further problems

9.1   *Problem 1 – Checking a bank statement*

Table 2.15   A bank statement

Payments	Receipts	Balance
£	£	£
		6.45
20.00		13.55 DR
	50.00	36.45
20.00		16.45
0.70		15.75
	100.00	115.75
10.00		105.75
23.58		82.17

Check the statement shown in Table 2.15. The keying sequence required to check the sequence of payments and receipts is shown in Table A1.

9.2   *Problem 2 – Updating a bank statement*

The following receipts and payments were subsequently added to the statement shown in Table 2.15. What is the revised balance?

> Received £5.45
> Paid   £8.79
> Paid   £6.99
> Paid   £10.83
> Paid   £15.00
> Paid   £9.75
> Received £12.75

The answer is given in Appendix B.

9.3 *Problem 3 – Checking a telephone bill*

Table 2.16   A Telephone bill

	£
Quarterly rental	8.65
130 units at 3p	3.90
Total	12.55
VAT at 8.00%	1.00
Total payable	13.55

Check the telephone bill shown in Table 2.16. The keying sequence for this problem is shown in Table A2.

9.4 *Problem 4 – Coping with increased charges*

The Post Office intend to raise the quarterly telephone rental charge to £9.50 and the government are altering the rate of VAT to 10%. By what proportion of time should the use of the telephone be reduced to keep the next bill to the same total as that in Table 2.16? The answer to this is given in Appendix B.

9.5 *Problem 5 – Calculating km/l*

I can travel 157 km on £3.00 of petrol bought from my local garage. The 'normal' pump price is 90p per 5 litres, but it is sold at 7p per 5 litres discount for cash. What km/l am I obtaining? The keying sequence is shown in Table A3.

9.6 *Problem 6 – Effect of discounts*

The garage mentioned in the previous problem changes ownership and the following method of serving is adopted. The first £3.00 of petrol is put in at 90p per 5 litres. The discount due is then calculated from the total number of litres used at 7p per 5 litres.

The amount of petrol this discount represents (at 90p per 5 litres) is then added. How much less petrol is obtained compared with the previous problem? The answer is given in Appendix B.

9.7   *Problem 7 – Calculating the number of floor tiles for a given area*

How many vinyl floor tiles, size 25 cm × 25 cm would be needed to cover a kitchen measuring 3.4 m × 2.6 m? The remains of cut tiles cannot be used. The keying sequence for this problem is given in Table A4.

3 Errors and accuracy

1 Introduction

The accuracy of any calculation performed on a calculator will depend on a number of factors. If care is not taken, it is easy to enter data incorrectly or to depress the wrong function key. The calculator used will itself have limitations with respect to the size of the numbers it can handle, and its operating accuracy will vary according to the functions and values used.

Errors due to rounding or truncation may be introduced when some of the decimal digits of a number are dropped. This may happen at different stages of the calculation, either when rounded or truncated numbers are entered, or when the calculator, through its basic logic, rounds or truncates a number.

The original data may only be approximate, particularly if they are concerned with physical measurements. In addition, the formulae used to obtain the required result may not be completely precise; for example, the formula may have been modified to ease the calculation, or some terms may have been dropped.

It is important, therefore, to understand the nature of the errors that are possible, how to prevent, detect and correct them, and the final result should be quoted to an accuracy which is known, and has practical significance.

2 Sources of error

2.1 *Incorrect entries*

These can be a common source of error, and can be prevented and detected by laying out the required calculation in the form of a keying table, keying in carefully and checking the display, noting some intermediate results for subsequent checking, and comparing the final result with an approximate value calculated roughly as a check that the result is of the right order.

For example,

$$\frac{2.6 \times 15.8}{40.8 - 31.2} \text{ is approximately} = \frac{3 \times 16}{10} = 4.8.$$

The actual calculated value is 4.279, which is of the right order. If a mistake was made and $2.6 + 15.8$ was calculated, instead of 2.6×15.8, the result would be 1.916, which is obviously wrong.

However, a small error in keying in a number would not be detected by this rough approximation. Keying in 40.5 instead of 40.8 gives a result of 4.417, which looks correct but is wrong. This error should be picked up when the calculation is repeated, as a check, preferably being performed in a different sequence.

Table 3.1 summarises the types of incorrect entries, and indicates how they can be detected and corrected.

2.2   *Calculation range*

The numbers that can be entered into a calculator or displayed as a result of a calculation must be within the calculation range of the machine.

For instance, if the calculator has the range 1×10^{-99} to $9 \times 10^{+99}$, then the calculation 2 EXP 3 × 1 EXP 99 i.e. $2 \times 10^3 \times 10^{99} = 2 \times 10^{102}$ (see also p. 42), which has a result greater than $9 \times 10^{+99}$, will produce an error; and this will be indicated in the display. No further calculations can be carried out until this error has been cleared. Similarly, the result of the calculation

$$6 \text{ EXP} -12 \div 2 \text{ EXP } 98 = (3 \times 10^{-110})$$

is a number which is too small to be held in the machine, and an error will be indicated when the = sign is depressed.

The error may be caused by inadvertently depressing the wrong function key. For instance,

$$5 \text{ EXP } 96 \times 4 \text{ EXP } 97 = (2 \times 10^{194})$$

Table 3.1 Detecting and correcting incorrect entries

Incorrect entry	Detected by	Corrected by
(a) Wrong digit key depressed.	Wrong number in display.	Clear display and re-enter if clear key available. Otherwise reverse operation, e.g. $+6$ entered instead of $+9$, then enter -6, followed by $+9$.
(b) Wrong function key depressed.	May not be detected until intermediate or final result checked.	(i) If noticed immediately before the next digit key is depressed, then entering the correct operation should eliminate the effect of the wrong entry. (ii) If the next number has been entered then a reversal of the whole operation, as in (a) above, will return the calculation to the stage it was before the wrong entry. (iii) If the error is not detected until an intermediate or final result is checked, then the last correct intermediate entry may be entered, but great care must be taken that it is correct and complete (not truncated or rounded).
c) Key to obtain dual function depressed when not required.	Indicator for dual function displayed.	Usually pressing this key again will eliminate the effect without affecting the previous calculation, so long as a function key has not been depressed. If it has, then reverse the operation. (If $\sqrt{x}$ depressed accidentally, x^2 will restore the situation.)
(d) Digit key depressed twice.	Display will show two digits of the same value and give the wrong number, e.g. 22. instead of 2.	Clear entry and re-enter correctly or proceed as in (a) above if no clear entry key is available.
(e) M+ or M− depressed twice.	The display will not show the effect which may not be noticed until the memory is recalled and used, or when a wrong intermediate or final result is found during checking.	If noticed immediately, the reverse operation can be applied (e.g. M+ twice followed by M−). Otherwise proceed as in (b(iii)).
(f) Digit key not depressed properly.	Digit not shown in display.	Press required digit key again, if noticed immediately. Otherwise, return to appropriate intermediate point and recommence calculation from there.
(g) Memory key not depressed properly.	If memory has not been used, then 'memory in use' indicator will not be displayed. If memory has been used then error may not be detected until an intermediate or final result has been checked. If unsure, check contents of memory with MEX, and press again to restore.	Press required memory key if next number has not been entered. Otherwise, return to appropriate intermediate point and recommence calculation from there.

will produce an error, when the intended calculation was

$$5 \text{ EXP } 96 + 4 \text{ EXP } 97 = (4.5 \times 10^{97})$$

which lies within the range of the calculator. Careful keying will prevent this type of error.

The range of numbers that can be used applies to the simple arithmetic functions $(+, -, \times, \div)$. Other functions may only be used with a restricted range of numbers which make the operation meaningful. For example, $\sqrt{x}$ may not be used with negative numbers; the sine of an angle cannot be greater than 1, therefore $\sin^{-1}$ of a number greater than 1 will produce an error.

2.3 *Operating accuracy*

The operating accuracy depends on the function used and on the number on which the operation is to be carried out (the argument). The accuracy is usually quoted in the manufacturer's instruction booklet as the maximum mantissa error.

Note: If the reader is not familiar with logarithms, then reference should be made to section 5 on p. 45 before considering the following example.

For example, the natural logarithm of a number (function *ln*) could on one calculator have an accuracy of 1 count in the 4th digit and on another calculator 1 count in the 10th digit. The difference is demonstrated using the calculation shown in Table 4.7 (p. 48). Using a calculator whose *ln* function was quoted as having an accuracy of 1 count in the 4th digit, 2.54 *ln* gave 0.93214, whereas using another calculator with an accuracy quoted as 1 count in the 10th digit, 2.54 *ln* gave 0.932164. Multiplying by 60 and taking the antilog (e^x) with the same accuracy, gave results of 1.9471×10^{24} and 1.9499478×10^{24} respectively, for the evaluation of 2.54^{60} (see also section 5.7, p. 47).

The maximum mantissa error is quoted, for some functions, over different ranges of the argument. For example, the tangent of an angle may have an accuracy of 4 counts in the 10th digit for angles between $0°$ and $89°$ and 1 count in the 6th digit for angles between $89°$ and $89.95°$.

A comparison of two calculators with different maximum mantissa errors gave the results show in Table 3.2.

Table 3.2   Values of tan from two calculators

angle	tan (calculator 1)	tan (calculator 2)
89°	57.289	57.289961
89.99°	5729.3	5729.5775
89.9999°	571420.	572953.63

2.4   Rounding errors

A calculator has a given number of digits in the display, but more digits may be used internally in the calculation. Thus, the operating accuracy may be quoted to the 10th digit, but the display may only show 8 of these digits. The number in the display has been truncated, i.e. the last two digits have been 'dropped'. The operating accuracy will only be affected if this truncated number is re-entered at some stage, when it will not be so accurate as its original internal representation.

Numbers may be rounded by the user before, during or after the calculation has been performed. The method normally used is to round up, that is, if a number is required to n decimal places, then 1 is added to the nth digit after the decimal point if the $(n + 1)^{\text{th}}$ digit is 5 or greater.

For example,

$$.396 \times .871 = .344916$$

The result is 0.34 rounded to two decimal places.

If we round the numbers before multiplying, i.e.

$$.40 \times .87$$

we obtain .348 = .35 rounded to two decimal places.

The effect of rounding errors needs to be known for addition, subtraction, multiplication, division, roots, powers and functions, so that the effect on calculations involving combinations of these operations can be calculated. Detailed error analysis is outside the scope of this book.

2.5   *Errors in the original data*

The accuracy of the original data and of the formulae used should be taken into account when estimating the accuracy of the result of the calculation.

The precision of physical measurements is better signified by calculating the relative or percentage error than the absolute error.

Note: The magnitude of the absolute error is equal to:

$$\text{calculated value} - \text{correct value}$$

The relative error is approximately equal to:

$$\frac{\text{calculated value} - \text{correct value}}{\text{calculated value}}$$

The percentage error is 100 × relative error.

For example, a measurement of the diameter of a small particle may be accurate to

$$.0016 \pm .0002 \text{ cm}$$

giving a small absolute error (.0002 cm) but a large percentage error of $(\frac{2}{16} \times 100) = 12.5\%$.

A distance measured during a survey of 4387 metres to the nearest 10 metres (large absolute error), has a small percentage error in the measurement of $\left(\frac{10}{4387} \times 100\right) = 0.23\%$.

4 Mathematical background of scientific functions

1 Introduction

Most scientific calculators have a common set of scientific function keys in addition to the four basic function keys (add, subtract, multiply and divide). Some calculators have extra function keys to make keying-in easier by reducing the number of key depressions. Other calculators have some of the function keys omitted because of design criteria (size of keyboard, circuitry). This section explains the purpose of some of the different function keys and how functions may be derived if the particular function keys are not available. In addition, the examples will give a better understanding of the mathematical meaning of the functions.

2 Raising a number to a power (y^x)

Raising a number to a power means multiplying the number (y) by itself for the specified number of times (x); for example,

$$y^3 = y \times y \times y.$$

x is called an index and further properties of indices are given in section 3.

If x is 2, i.e. $y \times y$ is required, then the answer is the *square* of the number and may be found by using the x^2 function key.

If a y^x function key is available then the keying sequence is:

(a) enter the number whose power is required (y)
(b) press y^x
(c) enter the index (x)
(d) press = or continue

Note: y need not be a single number but may be the result of evaluating an arithmetic expression consisting of several numbers and operations.

2.1 *Example 1*

Evaluate $((3.6 + 5.7) \times 2.8)^5$

See Table 4.1 for the keying sequence using the y^x key. Table 4.2 shows the keying sequence using multiplication only for raising to the power 5.

Table 4.1   Example 1a

Keying	Display	Comments
3.6		
+		
5.7		
×	9.3	
2.8		
=	26.04	
y^x		
5		Raise to 5th power
=	11973052.	$(26.04)^5$

Table 4.2   Example 1b

Keying	Display	Comments
3.6		
+		
5.7		
×	9.3	
2.8		
=	26.04	y
M+		Store in memory
×		
MR	26.04	Recall memory
×	678.08	y^2
MR	26.04	
×	17657.2	y^3
MR	26.04	
×	459794.6	y^4
MR	26.04	
=	11973052.	y^5

The answer is 11973052.

2.2 *Using the constant facility*

This facility may be used for repeated multiplication to give the required power as shown in Table 4.3. This method reduces the number of key depressions required, compared with the straightforward repeated multiplication method as shown in Table 4.2. However, a large number of depressions are still required for high powers; for example, y^{50} would, using this method, entail depressing the $=$ key 49 times (to obtain y^2 to y^{50}), and it can easily be depressed the wrong number of times.

Table 4.3   Example 1c

Keying	Display	Comments
3.6		
+		
5.7		
×		
2.8		
=	26.04	y
×		
=	678.08	y^2
=	17657.2	y^3
=	459794.6	y^4
=	11973052.	y^5

3   Laws of indices

At this stage it is useful to consider the laws of indices, as these can be used to simplify expressions for use on the calculator.

3.1   *Multiplication*

$$y^5 = y \times y \times y \times y \times y$$

and

$$y^3 = y \times y \times y$$

therefore,

$$y^8 = y^5 \times y^3$$

or generally,

$$y^a \times y^b = y^{a+b} \tag{4.1}$$

3.2   *Division*

From 3.1,

$$y^5 \div y^3 = y^2$$

or generally,

$$y^a \div y^b = y^{a-b} \tag{4.2}$$

3.3   *Powers*

$$(y^5)^3 \text{ means } y^5 \times y^5 \times y^5$$
$$= y^{5+5+5} = y^{15}$$

or generally,

$$(y^a)^b = y^{ab} \tag{4.3}$$

3.4 y^0

Let us now consider the meaning of y^0; this is equal to

$$y^{x-x} = y^x \div y^x$$

i.e. $$y^0 = 1 \tag{4.4}$$

thus, any number raised to the power 0 is equal to 1 (definition).

3.5 *Negative indices*

The index may be negative which has the following effect:

$$y^{-x} \times y^x = y^{-x+x} = y^0 = 1$$

i.e.
$$y^{-x} \times y^x = 1 \qquad (4.5)$$

Dividing each side of (4.5) by y^x,

$$y^{-x} = \frac{1}{y^x} \qquad (4.6)$$

that is, y^{-x} is the reciprocal of y^x.

3.6 *Fractional indices*

The same rules apply to fractional indices, therefore:

$$y^{\frac{1}{2}} \times y^{\frac{1}{2}} = y^{\frac{1}{2}+\frac{1}{2}} = y^1 = y$$

This means that $y^{\frac{1}{2}}$ must represent the square root of y, that is,

$$y^{\frac{1}{2}} = \sqrt{y}$$

Also,

$$y^{\frac{1}{3}} \times y^{\frac{1}{3}} \times y^{\frac{1}{3}} = y$$

that is,

$$y^{\frac{1}{3}} = \sqrt[3]{y} \text{ (the cube root of } y)$$

or generally,

$$y^{1/x} = \sqrt[x]{y} \qquad (4.7)$$

See also *Teach Yourself Algebra* (P. Abbott).

4 Use of nested parentheses

The laws of indices may be used to change an expression in order to eliminate high powers. This is done by first determining the prime factors of the power, and using nested parentheses (i.e. brackets within brackets).

Note: A prime factor is a number which cannot be divided by any other number than itself and unity (1).

It is particularly convenient to change the expression so that only powers of 2 (i.e. x^2) need to be used.

4.1 *Example 2*

Evaluate 2.54^{60} without using the y^x facility.

First convert 2.54^{60} using nested parentheses.

i.e.
$$(2.54)^{60} = (2.54)^{64-4}$$
$$= \frac{(2.54)^{64}}{(2.54)^4} = \frac{(((((((2.54)^2)^2)^2)^2)^2)^2)^2}{((2.54)^2)^2}$$

Table 4.4 shows the keying sequence using the x^2 facility. Alternatively, the constant facility may be used to obtain the squares. The answer is 1.9499×10^{24}.

Table 4.4 Example 2

Keying	Display		Comments
2.54			
x^2	6.4516		
x^2	41.623		
M+			$(2.54)^4$
x^2	1732.4		
x^2	3001507.7		
x^2	9.0090	12	
x^2	8.1162	25	$(2.54)^{64}$
÷			
MR	41.623		
=	1.9499	24	$(2.54)^{60}$

If the power is odd, for example, 2.54^{61}, then, we have

$$(2.54)^{61} = (2.54)^{64-4+1}$$

$$= \frac{(2.54)^{64}}{(2.54)^4} \times 2.54$$

i.e. the answer previously obtained has to be multiplied by 2.54 once, which gives 4.9527×10^{24} as the evaluation of 2.54^{61}.

5 Logarithms

5.1 *Multiplication and division*

If numbers can be expressed as powers of a constant number (base), e.g. 10, then multiplication of the number only involves adding the indices.

For example,

$$5.6235 \times 31.623 = 10^{0.75} \times 10^{1.5}$$
$$= 10^{0.75+1.5}$$
$$= 10^{2.25}$$
$$= 177.83$$

Therefore, $177.83 = 10^{2.25}$ and
2.25 is called the logarithm of 177.83 to base 10.

In a similar way, two numbers may be divided by subtracting the logarithms of the numbers.

For example,

$$17.783 \div 5.6235 = 10^{1.25-0.75}$$
$$= 10^{0.5}$$
$$= 3.1623$$

5.2 *Logarithm of a power and of a root*

Two further rules need to be known. The logarithm of a power of a number is given by:

$$\log y^x = x \log y \qquad (4.8)$$

that is, the log of the number is multiplied by the power.

The xth root of a number is the log of the number divided by x, i.e.

$$\log \sqrt[x]{y} = \log (y^{1/x}) \tag{4.9}$$

$$= \frac{1}{x} \log y$$

$$= (\log y) \div x$$

5.3   Antilogarithms

In 5.1, 2.25 was shown to be the logarithm of 177.83 to base 10. Reversing the procedure, we say that the antilogarithm of 2.25 to base 10 is 177.83, i.e. the number whose logarithm to base 10 is 2.25, is 177.83.

5.4   Naperian or natural logarithms

So far only logarithms to base 10 have been considered. Logarithms to base e are known as Naperian or natural logarithms. e has the value 2.7182818 to 8 significant figures.

5.5   Changing the base of logarithms

Natural logarithms are used in many scientific calculations. However, if logarithms to base 10 are required, and a $\log_{10} x$ key is not available, then logarithms to base e can be found ($ln\ x$ or $\log_e x$ key) and converted to base 10 as follows:

(a)   enter number whose $\log_{10}$ is required
(b)   press ln key
(c)   divide by 2.302585

Conversely, given the log of a number to base 10, the number may be found by multiplying by 2.302585 and finding e^x of the result.

5.6 *Example 3*

(a) Find $\log_{10} 4.157$, without using the $\log_{10} x$ key.
See Table 4.5 for the keying sequence.

Table 4.5   Example 3a

Keying	Display	Comments
4.157		
ln	1.4247	$\log_e 4.157$
÷		
2.302585		Conversion factor for base 10
=	0.61878	$\log_{10} 4.157$

The answer is 0.61878.

(b) Find the antilog to base 10 of the result of (a).
This should give you the original number.
See Table 4.6 for the keying sequence.

Table 4.6   Example 3b

Keying	Display	Comments
0.61878		$\log_{10} 4.157$
×		
2.302585		$\log_e 10$
=	1.4247	
e^x	4.157	

5.7 *Example 4*

Use logarithms to find 2.54^{60} and compare your answer with that found in section 4.1. The keying sequence is shown in Table 4.7.

Table 4.7   Example 4

Keying	Display	Comments
2.54		
ln	0.93216	$\log_e$ 2.54
×		
60		Raise to 60th power
=	55.929	
e^x	1.9499 24	$2.54^{60} = 1.9499 \times 10^{24}$

The answer is 1.9499×10^{24}.

Note on accuracy: see section 2.3, p. 36.

6   Finding the square root of a number ($\sqrt{x}$)

6.1   *Using the $\sqrt{x}$ key*

The square root of a number may be found easily using the $\sqrt{x}$ key, by entering the number followed by a depression of the $\sqrt{x}$ key. If the square root of an expression is required, the = key should be depressed before the $\sqrt{x}$ key.

6.2   *Using logarithms*

The logarithm of the square root of a number is equal to the logarithm of the number divided by 2 (see section 5.2).
    Thus,

$$\log \sqrt{15} = \tfrac{1}{2} \log 15$$

(a)   enter 15
(b)   press *ln*
(c)   divide by 2 and press =
(d)   press e^x to obtain $\sqrt{15}$

The answer is 3.87298.

7 Finding the xth root of a number ($^x\sqrt{y}$)

7.1 Using the $^x\sqrt{y}$ key

If a $^x\sqrt{y}$ key is available then the keying sequence is:

(a) enter the number (y) whose xth root is required
(b) press the $^x\sqrt{y}$ key
(c) enter x, then =

7.2 Using logarithms

From section 5.2, we have:

$$\log {}^x\sqrt{y} = (\log y) \div x.$$

7.3 Example 5

Find $^5\sqrt{6.12}$ using logarithms.
The keying sequence is shown in Table 4.8.

Table 4.8   Example 5

Keying	Display	Comments
6.12		
ln	1.8115	$\log_e 6.12$
÷		Find 5th root
5		
=	0.36231	
e^x	1.4366	$^5\sqrt{6.12} = 1.4366$
×		
=	2.0639	to check answer use
=	2.9651	constant facility,
=	4.2599	or use y^x
=	6.1199	

The answer is 1.4366.

8   Exponential series

The exponential series below has been found to be convergent for all values of x.

$$e^x = 1 + x + \frac{x^2}{2!} + \frac{x^3}{3!} + \ldots \rightarrow \infty \qquad (4.10)$$

Note:

$$2! \text{ (factorial 2)} = 1 \times 2$$

$$3! \text{ (factorial 3)} = 1 \times 2 \times 3$$

The accuracy of the evaluation is determined by the number of terms used. The evaluation will be as follows for 6 terms:

$$e^x = 1 + \frac{x^2}{2} + \frac{x^3}{6} + \frac{x^4}{24} + \frac{x^5}{120} + \frac{x^6}{720} \qquad (4.11)$$

If only four functions $(+, -, \times, \div)$ are available on the calculator, the tediousness of this calculation can be reduced by restating the expression in nested parenthical form as follows (see p. 44):

$$e^x = 1 + x\left(1 + \frac{x}{2}\left(1 + \frac{x}{3}\left(1 + \frac{x}{4}\left(1 + \frac{x}{5}\left(1 + \frac{x}{6}\right)\right)\right)\right)\right) \qquad (4.12)$$

Remember always to start an evaluation with the innermost brackets and work outwards.

To obtain the value of e, put $x = 1$ in (4.12).

The keying sequence is as follows (starting with the innermost brackets):

$$1 \div 6 + 1 \div 5 + 1 \div 4 + 1 \div 3 + 1 \div 2 + 2 = 2.7180555$$

Using 7 and 8 terms gives values for e of 2.7182539 and 2.7182787 respectively.

The value of e may also be found by keying in 1 followed by e^x. The accuracy of this will vary as specified in the instruction booklet supplied by the calculator manufacturer. A calculator, for which the accuracy of e^x is given as 1 count in the 10th digit, displays e (8 digit display) as 2.7182818.

9 Trigonometric functions

The trigonometric functions sin, cos, tan, and their inverse functions $\sin^{-1}$ (arcsine), $\cos^{-1}$ (arcosine), $\tan^{-1}$ (arctangent) and π are described in chapter 5.

10 Hyperbolic functions

Trigonometric functions are connected with the circle. In a similar way, certain functions of e^x are connected with the hyperbola and are known as the hyperbolic functions.

10.1 *Hyperbolic sine of x (sinh x)*

$$\sinh x = \tfrac{1}{2}(e^x - e^{-x}) \tag{4.13}$$

Some calculators have a sinh x function key, alternatively e^x may be used in (4.13).

Note: $$e^{-x} = \frac{1}{e^x}$$

The series for sinh x is found by considering the series for e^x and e^{-x}.

i.e. $$\sinh x = x + \frac{x^3}{3!} + \frac{x^5}{5!} + \ldots \to \infty \tag{4.14}$$

In nested parenthical form, using the first four items only,

$$\sinh x = x\left(1 + \frac{x^2}{6}\left(1 + \frac{x^2}{20}\left(1 + \frac{x^2}{42}\left(1 + \frac{x^2}{72}\right)\right)\right)\right) \tag{4.15}$$

10.1.1 *Example 6*

Find sinh 3 using

(a) e^x
(b) nested parentheses

The keying sequences are shown in Tables 4.9 and 4.10.

Table 4.9 Example 6a

Keying	Display	Comments
3	3.	
e^x	20.085	e^3
M+		
$\frac{1}{x}$	0.049787	e^{-3}
M−		
MR	20.035	$e^3 - e^{-3}$
÷		
2		
=	10.0179	sinh 3 = 10.0179

Table 4.10 Example 6b

Keying	Display	Comments
9		x^2
M+		
÷		
72		
+	0.125	
1		
×	1.125	
MR	9.	x^2
÷	10.125	
42		
+	0.24107	
1		
×	1.2410	
MR	9.	x^2
÷	11.169	
20		
+	0.55848	
1		
×	1.5584	
MR	9.	x^2
÷	14.026	
6		
+	2.3377	
1		
×	3.3377	
3		x
=	10.013169	sinh 3 = 10.0132

The answers are (a) 10.0179, (b) 10.0132.

10.2 *Hyperbolic cosine of x (cosh x)*

$$\cosh x = \tfrac{1}{2}(e^x + e^{-x}) \tag{4.16}$$

$$= 1 + \frac{x^2}{2!} + \frac{x^4}{4!} + \cdots$$

i.e. $$\cosh x = \left(1 + \frac{x^2}{2}\left(1 + \frac{x^2}{12}\left(1 + \frac{x^2}{30}\left(1 + \frac{x^2}{56}\right)\right)\right)\right) \tag{4.17}$$

10.2.2 *Example 7*

Find cosh 4 using

(a) e^x
(b) nested parentheses (equation 4.17)

The keying sequence is shown in Tables 4.11 and 4.12.

Table 4.11 Example 7a

Keying	Display	Comments
4		
e^x	54.598	e^4
M+		
$\frac{1}{x}$	0.018315	e^{-4}
M+		
MR	54.616	$e^4 + e^{-4}$
$\div$		
2		
$=$	27.3082	cosh 4 = 27.3082

Table 4.12   Example 7b

Keying	Display	Comments
16		x^2
÷		
56		
+	0.28571	
1		
×	1.2857	
16		x^2
÷	20.571	
30		
+	0.68571	
1		
×	1.6857	
16		x^2
÷	26.971	
12		
+	2.2476	
1		
×	3.2476	
8		$x^2/2$
+	25.980	
1		
=	26.980	cosh 4 = 26.980

The answers are (*a*) 27.3082, (*b*) 26.980.

10.3   *Hyperbolic tangent of x (tanh x)*

$$\tanh x = \frac{e^x - e^{-x}}{e^x + e^{-x}} = \frac{\sinh x}{\cosh x} \qquad (4.18)$$

10.4   *Other identities*

$$\operatorname{cosech} x = \frac{1}{\sinh x} \qquad (4.19)$$

$$\text{sech } x = \frac{1}{\cosh x} \qquad (4.20)$$

$$\coth x = \frac{1}{\tanh x} = \frac{\cosh x}{\sinh x} \qquad (4.21)$$

$$\cosh x + \sinh x = e^x \qquad (4.22)$$

$$\cosh x - \sinh x = e^{-x} \qquad (4.23)$$

Multiplying (4.22) and (4.23) gives:

$$\cosh^2 x - \sinh^2 x = e^0 = 1 \qquad (4.24)$$

11   Inverse hyperbolic functions

$\text{Sinh}^{-1} x$, $\cosh^{-1} x$, $\tanh^{-1} x$, $\text{cosech}^{-1} x$, $\text{sech}^{-1} x$ and $\coth^{-1} x$ represent the angle (in radians) whose sinh, cosh, tanh, cosech, sech and coth have the value x.

$$\sinh^{-1} x = ln(x + \sqrt{x^2 + 1}) \qquad (4.25)$$

and

$$\cosh^{-1} x = ln(x \pm \sqrt{x^2 - 1}) = \pm ln(x + \sqrt{x^2 - 1}) \qquad (4.26)$$

Note: both signs are valid in this case.

$$\tanh^{-1} x = \tfrac{1}{2} ln\left(\frac{1+x}{1-x}\right) \qquad (4.27)$$

where $-1 < x < 1$.

11.4   *Example 8*

Evaluate $\sinh^{-1} 1.6$, by using (4.25).

The keying sequence is shown in Table 4.13.

Table 4.13   Example 8

Keying	Display	Comments
1.6		x
M+		
x^2	2.56	1.6^2
+		
1		
=	3.56	$1.6^2 + 1$
$\sqrt{x}$	1.8867	$\sqrt{1.6^2 + 1}$
+		
MR	1.6	x
=	3.4867	$1.6 + \sqrt{1.6^2 + 1}$
ln	1.24898	$\sinh^{-1} 1.6 = 1.2490$

The answer is 1.2490.

12   Summary of functions used for scientific calculations

FUNCTION	ACTION
x^2	Computes the square of the number displayed.
y^x	Raises the first number entered (y) to the power of the second number entered (x).
$\sqrt{x}$	Computes the square root of the number displayed.
$\sqrt[x]{y}$	Finds the xth root of y, where y and x are the first and second numbers respectively.
$1/x$	Computes the reciprocal of the number displayed.
$\log x$	Computes the logarithm to base 10 of the number displayed.
$\ln x$	Computes the logarithm to base e (natural log) of the number displayed.
e^x	Computes the antilog of the displayed number to base e (raises e to the displayed power).
sin	Computes the sine of the angle displayed.
cos	Computes the cosine of the angle displayed.

tan	Computes the tangent of the angle displayed.
$\begin{cases} \sin^{-1} \text{ (arcsine)} \\ \cos^{-1} \text{ (arcos)} \\ \tan^{-1} \text{ (arctan)} \end{cases}$	Computes the inverse trigonometric functions (ARC) i.e. finds the angle whose sin, cos or tan is the value displayed.
π	Enters the value of π into the display.

5   Trigonometric functions

1   The right-angled triangle

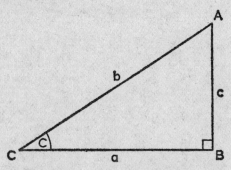

Figure 5.1   A right-angled triangle

Figure 5.1 shows a right-angled triangle ABC of sides a, b and c, in which angle B is equal to $90°$ (i.e. it is a right-angle). The trigonometric ratios for angle C are defined as follows:

$$\sin C = \frac{c}{b} \tag{5.1}$$

$$\cos C = \frac{a}{b} \tag{5.2}$$

$$\tan C = \frac{c}{a} = \frac{c}{b} \cdot \frac{b}{a} = \frac{\sin C}{\cos C} \tag{5.3}$$

In addition, for any angle x,

$$\frac{1}{\cos x} = \sec x \qquad \text{(secant)} \tag{5.4}$$

$$\frac{1}{\sin x} = \operatorname{cosec} x \qquad \text{(cosecant)} \tag{5.5}$$

$$\frac{1}{\tan x} = \cot x \qquad \text{(cotangent)} \qquad (5.6)$$

Note also that,

$$\frac{c}{b} = \cos A = \cos (90° - C)$$

i.e.
$$\sin C = \cos (90° - C) \qquad (5.7)$$

Similarly,
$$\cos C = \sin (90° - C) \qquad (5.8)$$
$$\tan C = \cot (90° - C) \qquad (5.9)$$

1.1 *Example 1*

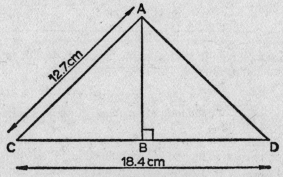

Figure 5.2 Example 1

In Figure 5.2, $AC = AD = 12.7$ cm and $CD = 18.4$ cm. Find angle CAB ($C\widehat{A}B$).

Since triangle ACD has two sides equal (AC and AD), the line AB divides the line CD into two equal parts. Triangles ABC and ABD are said to be congruent, since the corresponding angles and sides of each are the same.

$$\sin C\widehat{A}B = \frac{9.2}{12.7}$$

and
$$C\widehat{A}B = \sin^{-1}\left(\frac{9.2}{12.7}\right)$$

The required angle is found by using the inverse sine function (arcsine). Using a calculator, the keying sequence is as shown in Table 5.1.

Table 5.1 Example 1

Keying	Display	Comments
9.2		
÷	9.2	
12.7		
=	0.72440	sin $C\widehat{A}B$
sin⁻¹	46.419	46.42°
−		
46		Find decimal part
×	0.41974	
60		Convert to minutes
=	25.184	25′

The answer given in decimals of a degree is 46.42°, or 46° 25′ where 1° = 60′ (minutes).

1.2 Problem 1 – Calculating the distance between landmarks

Two landmarks (X, Y) on a mountain slope have been found to be 428.6 metres apart. The line XY is inclined at 15° 47′ to the horizontal. Find the horizontal and vertical distances between the two landmarks.

Let the horizontal distance be x metres and the vertical distance be h metres, then,

$$x = \cos 15°\ 47′ \times 428.6$$
$$h = \sin 15°\ 47′ \times 428.6$$

The keying sequence is shown in Table A5.

The horizontal distance is 412.44 metres, and the vertical distance is 116.57 metres.

1.3 *Problem 2 – Testing the stability of a lorry*

The stability of a lorry may be tested by tilting it through a given angle. The centre of gravity of a lorry is at a vertical height of 1.62 m above the ground on a level road, and the distance between the nearside and offside wheels is 2.15 m. Determine whether the lorry will remain stable, i.e. will not topple over when it is tilted at an angle of 30° to the horizontal.

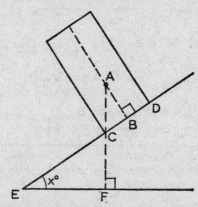

Figure 5.3 Testing the stability of a lorry

The lorry will become unstable when a vertical line through its centre of gravity (*A* in Figure 5.3) does not fall within its base (*DC*), i.e. $C\widehat{A}B$ is greater than

$$\tan^{-1}\frac{BC}{AB} = \tan^{-1}\left(\frac{2.15}{2} \times \frac{1}{1.62}\right)$$

But $A\widehat{C}B = E\widehat{C}F$ (equal opposite angles).

Therefore, $C\widehat{E}F = C\widehat{A}B = x$

The keying sequence is shown in Table A6.

The lorry will topple over when it is inclined at an angle of 33° 34′ to the horizontal, therefore it will remain stable when it is inclined at an angle of 30°.

2   Triangles not containing a right angle

2.1   *The sine rule*

The sine rule gives a relationship between the sides (a, b, c) and the sines of the angles (A, B, C) of a triangle, as follows:

$$\frac{a}{\sin A} = \frac{b}{\sin B} = \frac{c}{\sin C} \tag{5.10}$$

2.2   *Example 2*

Given $B = 62°\ 43'$ and $C = 30°\ 16'$ and side $C = 24.7$ cm, use the sine rule to find side b in cm.

$$b = \frac{\sin 62°\ 43' \times 24.7}{\sin 30°\ 16'}$$

See Table 5.2 for the keying sequence.

Table 5.2   Example 2

Keying	Display	Comments
16		16'
÷	16.	
60		Convert to decimals of a degree
+	0.26666	
30		Add 30°
=	30.266	
sin	0.50402	sin 30° 16'
M+		Store in memory
43		43'
÷	43.	
60		Convert to decimals of a degree
+	0.71666	
62		Add 62°
=	62.716	
sin	0.88875	sin 62° 43'
×		
24.7		Multiply by c
÷	21.952	
MR	0.50402	Divide by sin 30° 16'
=	43.553	$b = 43.6$ cm

The answer is 43.6 cm.

2.3 *The cosine rule*

The cosine rule gives a relationship between the sides of a triangle and the cosine of one angle. For example,

$$c^2 = a^2 + b^2 - 2ab \cos C \qquad (5.11)$$

2.4 *Example 3*

Given $a = 18.0$ cm, $b = 13.5$ cm and $c = 7.5$ cm, find the angle C in degrees and minutes. Use the cosine rule in the following form:

$$\cos C = \frac{a^2 + b^2 - c^2}{2ab} = \frac{18^2 + 13.5^2 - 7.5^2}{2 \times 18 \times 13.5}$$

The keying sequence is given in Table 5.3.

Table 5.3   Example 3

Keying	Display	Comments
18		a
x^2	324.	18^2
+		
13.5		b
x^2	182.25	13.5^2
−		
7.5		c
x^2	56.25	7.5^2
÷	450.	$18^2 + 13.5^2 - 7.5^2$
2		
÷	225.	
18		a
÷	12.5	
13.5		b
=	0.92592	$\cos C$
$\cos^{-1}$	22.191	$C = 22.191°$
−		
22		
×	0.19160	
60		Convert to minutes
=	11.496	11

The answer is 22° 11′.

3   Angles greater than 90°

So far the angles that have been used have been 90° or less. For angles between 90° and 180°, the sine of the angle is equal to the sine of (180° − the angle), i.e.

$$\sin x = \sin(180° - x) \text{ for } 90° < x < 180°$$

The values of the cosines of angles between 90° and 180° can be expressed as,

$$\cos x = -\cos(180° - x) \text{ for } 90° < x < 180°$$

Table 5.4 shows the trigonometric ratios of angles in the four quadrants.

Table 5.4   Sine, cosine and tangent of angles of any magnitude

$0° < \theta < 90°$	$90° < \theta < 180°$	$180° < \theta < 270°$	$270° < \theta < 360°$
$\sin \theta$	$\sin(180° - \theta)$	$-\sin(\theta - 180°)$	$-\sin(360° - \theta)$
$\cos \theta$	$-\cos(180° - \theta)$	$-\cos(\theta - 180°)$	$\cos(360° - \theta)$
$\tan \theta$	$-\tan(180° - \theta)$	$\tan(\theta - 180°)$	$-\tan(360° - \theta)$

4   Solution of triangles

Using the formulae previously stated, it is now possible to find the values of all the sides and angles of a triangle, i.e. 'solve' the triangle, when only some of them are given.

4.1   *Given three sides*

The cosine rule may be used to find the three angles independently. The result can then be checked by testing that the sum of the three angles is 180°.

See Table 5.3 for the keying sequence.

4.2 *Given two sides and an included angle*

The cosine rule again may be used for this case. For example, if sides b and c and angle A are given, then the third side is given by

$$a = \sqrt{b^2 + c^2 - 2bc \cos A} \qquad (5.12)$$

4.3 *Problem 3 – Using the cosine rule*

Given $A = 96°$, $b = 6.9$ cm, $c = 23.1$ cm, use the cosine rule to find a in cm.

The answer is given in Appendix B.

4.4 *Given all the angles and one side*

In this case, the most appropriate formula to use is the sine rule (see Table 5.2 for the keying sequence).

4.5 *Given two sides and an angle opposite to one of the sides*

The triangle can be solved by using the sine rule. However, there are two solutions to the problem when the side opposite to the given angle is less than the side adjacent to the angle, and in addition is greater than the sine of the angle multiplied by the adjacent side.

4.6 *Problem 4 – Using the sine rule*

Given a triangle ABC in which $AB = 53.6$ cm, $AC = 49.2$ cm and $A\widehat{B}C = 35°$, find $B\widehat{C}A$.

Note: As AC is less than AB, there will be two possible solutions if $AC > AB \sin A\widehat{B}C$.

The answer is given in Appendix B.

5 Area of a triangle

The area of a triangle is equal to:

$$\tfrac{1}{2} \times \text{the base} \times \text{perpendicular height}$$

The area of triangle ABC may be found by using

$$\tfrac{1}{2}ab \sin C \quad \text{or} \quad \tfrac{1}{2}bc \sin A \quad \text{or} \quad \tfrac{1}{2}ac \sin B$$

Alternatively, the area may be expressed in terms of the three sides.

Let the sum of the three sides $(a + b + c) = 2s$, then the area of

$$\triangle ABC = \sqrt{s(s - a)(s - b)(s - c)} \qquad (5.13)$$

5.1 *Example 4*

Given that a triangle has sides $a = 6.0$ cm, $b = 3.6$ cm and $c = 5.3$ cm, find the area of the triangle using these values only.

The keying sequence is shown in Table 5.5.

The area of the triangle is 9.46 cm^2.

Table 5.5 Example 4

Keying	Display	Comments
6		a
+		
3.6		b
+	9.6	
5.3		c
÷	14.9	
2		
=	7.45	s
M+		
−		
6		
=	1.45	$s - a$

Table 5.5—*continued*

Keying	Display	Comments
MEX	7.45	
—		
3.6		
×	3.85	$s - b$
MR	1.45	
=	5.5825	$(s - a)(s - b)$
MEX		
7.45		s
—		
5.3		
×	2.15	$s - c$
MR	5.5825	
×	12.002	$(s - a)(s - b)(s - c)$
7.45		
=	89.417	
$\sqrt{x}$	9.4560	Area of triangle = 9.46 cm²

5.2   *Problem 5 – Finding the area of a field*

Figure 5.4 shows a plan of a field *PQRS*. The following measurements were taken when the field was surveyed.

$$PR = 237 \text{ metres}, \ Q\widehat{P}R = 29°, \ Q\widehat{R}P = 42°, \ R\widehat{P}S = 73°$$
$$\text{and } P\widehat{R}S = 38°.$$

Calculate the area of the field in m².

The area of the field is the sum of the areas of the triangles *PQR* and *PSR*.

Area of $\triangle PQR = \frac{1}{2} \times 237 \times QX$

$\qquad\qquad\qquad = \frac{1}{2} \times 237 \times \sin 42° \times QR$

$\qquad\qquad\qquad = \frac{1}{2} \times 237 \times \sin 42° \times \dfrac{237 \times \sin 29°}{\sin P\widehat{Q}R}$ (using the sine rule)

where $\sin P\widehat{Q}R = \sin (180° - 29° - 42°) = \sin 109° = \sin 71°$.

(See Table 5.4.)

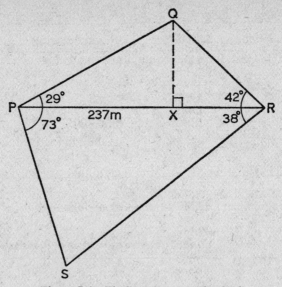

Figure 5.4   Finding the area of a field

Similarly,

$$\text{Area of } \triangle PSR = \tfrac{1}{2} \times 237 \times \sin 38° \times \frac{237 \times \sin 73°}{\sin \widehat{PSR}}$$

where $\sin PSR = \sin(180° - 73° - 38°) = \sin 69°$.

The keying sequence for finding the area of triangle PQR is shown in Table A7.

The areas of $\triangle PQR$ and $\triangle PSR$ are 9636 m and 17 711 m respectively, so that the area of the field is 27 347 m².

6   Radius of circumcircle of a triangle

The radius (r) of the circumcircle of a triangle ABC (sides a, b, c) and angles (A, B, C) is given by

$$r = \frac{a}{2 \sin A} = \frac{b}{2 \sin B} = \frac{c}{2 \sin C} \qquad (5.14)$$

6.1 Problem 6 – *Finding the radius of a circular track*

Find the radius of a circular track, which is to pass through points
P, *Q* and *R* which form a triangle, as shown in Figure 5.5.

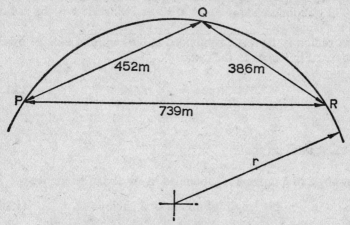

Figure 5.5 Finding the radius of a circular track

$PQ = 452$ metres, $QR = 386$ metres and $PR = 739$ metres.
Using the cosine rule,

$$\cos P\widehat{Q}R = \frac{452^2 + 386^2 - 739^2}{2 \times 452 \times 386}$$

$$\text{Radius of track} = \frac{PR}{2 \sin P\widehat{Q}R}$$

The keying sequence is shown in Table A8.

The radius of the track will be 443.3 metres.

7 Radian or circular measure

For all circles, it can be shown that

$$\frac{\text{circumference}}{\text{diameter }(d)} \text{ is constant } (= \pi)$$

The value of π cannot be expressed exactly as a fraction. A good approximation is $\frac{22}{7}$ (3.1428571) or better $\frac{355}{113}$ (3.1415929). The π key on a scientific calculator may give a value slightly different from the latter depending on its accuracy.

The circumference is equal to πd or $2\pi r$ where r is the radius of the circle.

A radian is the angle subtended at the centre of a circle by an arc equal in length to the radius.

$$1 \text{ degree} = \frac{\pi}{180} \text{ radians} \qquad (5.15)$$

8   Length of an arc

An angle of θ radians is subtended by an arc of length θr.

i.e.   the length of an arc of a circle $= r\theta$ $\qquad (5.16)$

9   Area of a circle, sector and segment

In Figure 5.6, the area of the circle $= \pi r^2$ $\qquad (5.17)$

and the area of the sector $= \frac{1}{2}r^2\theta$ $\qquad (5.18)$

where θ is in radians.

AB is a chord which divides the area of the circle into a minor segment (less than a semicircle, shaded portion in Figure 5.6) and a major segment (the rest of the circle).

9.1   *Example 5*

Find the area of the major segment of a circle of radius 4 cm, given the length of the chord $AB = 7.6$ cm.

First consider $\triangle AOB$, which is isosceles. A perpendicular

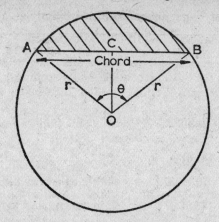

Figure 5.6 Area of a circle, sector and segment

from θ to AB will bisect AB and $\widehat{AOB}$ to give two congruent triangles OAC and OBC (see Figure 5.6).

Let $\theta = 2x$, then $\sin x = \dfrac{3.8}{4}$

and $\theta = 2 \sin^{-1} x$ radians (if set to work in radians)

or $\theta = (2 \sin^{-1} x) \times \dfrac{\pi}{180}$ radians (if the calculator is set to work in degrees).

The area of the major segment = area of circle − area of minor segment

$$= \pi r^2 - \tfrac{1}{2} r^2 (\theta - \sin \theta)$$

$$= \frac{r^2}{2}(2\pi - \theta + \sin \theta) \tag{5.19}$$

See Table 5.6 for the keying sequence.

Table 5.6   Example 5

Keying	Display	Comments
3.8		half chord
÷		
4		
=	0.95	sin x
sin^{-1}	71.805	
×		
2		
=	143.61	$\theta°$
M+		
sin	0.59327	sin θ (alternatively use sequence 180 − MR = sin)
MEX	143.61	$\theta°$ in display, sin θ in memory
×		
π	3.1415926	
÷		
180		
=	2.5064	θ radians
M−		
2		
×		
π	3.1415926	
=	6.2831852	2π
+		
MR	−1.91319	sin θ − θ
×	4.369988	
8		$r^2/2$ (by mental arithmetic)
=	34.9599	area of segment = 34.96 cm^2

The area of the major segment is 34.96 cm^2.

9.2   Problem 7 – Finding the area of triangle AOB

Given that the angle subtended at the centre of a circle of radius 10.25 cm is 1.9 radians, find the area of the triangle formed by the two radii and the chord joining them (see Figure 5.6).

Let $\theta = 1.9$ radians $= \dfrac{180 \times 1.9}{\pi}$ degrees

In Figure 5.6,

$$AC = \sin \frac{\theta}{2} \times r$$

$$OC = \cos \frac{\theta}{2} \times r$$

area of triangle $AOB = AC \times OC$.

The answer is given in Appendix B.

10   Small angles

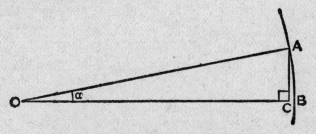

Figure 5.7   Calculation of small angles

In Figure 5.7 α is a small angle, such that, $AC \simeq AB$, and $OA \simeq OC$, where AB is the arc of a circle of radius OA.

Note: $\simeq$ means approximately equal to.

It can be shown that for small angles,

$$\tan \alpha \simeq \sin \alpha \simeq \alpha \qquad \qquad (5.20)$$

and $\qquad\qquad\qquad \cos \alpha \simeq 1 \qquad\qquad\qquad\qquad (5.21)$

10.1   *Problem 8 – Calculation of* 2α

In Figure 5.7, given $AC = 0.292$ cm and $OC = 12.9$ cm, find 2α (which is small).

$$\alpha \simeq \tan \alpha = \frac{AC}{OC}$$

therefore,

$$2\alpha = \frac{2 \times 0.292}{12.9}$$

The answer is given in Appendix B.

Further reading, examples and exercises on trigonometric relationships may be found in *Teach Yourself Trigonometry* (P. Abbott).

6 Volumes and surface areas

1 Introduction

The calculation of the volume of solid objects requires a knowledge of the cross-sectional area and the height of the solid. If the cross-sectional area is uniform then the volume is equal to:

cross-sectional area × perpendicular height.

If the cross-sectional area is not uniform, then different formulae will apply according to the shape of the object.

The surface area of any solid is found by adding together the face areas and the curved surface areas.

See also *Teach Yourself Mathematics* (L. C. Pascoe) and *Teach Yourself Geometry* (P. Abbott).

2 The cuboid

The cuboid has a rectangular cross-sectional area $= l \times w$, where l is the length and w the width. If l is equal to w then the cross-sectional area is a square.

$$\text{The volume of a cuboid} = l \times w \times h \qquad (6.1)$$

If these three dimensions are expressed in cm then the unit of volume is cm^3.

The surface area is the sum of the face areas, i.e.

$$2(lw + lh + wh) \qquad (6.2)$$

2.1   *Problem 1 – Finding the length and width of a cuboid*

A cuboid of square cross-section has a volume (V) of 67.3 cm³, and a height (h) of 4.2 cm. Find its length (l) and width (w).

The answer is given in Appendix B.

3   The cylinder

The cross-sectional area of a cylinder is a circle of area πr^2, where r is the radius of the circle.
   Therefore,

$$\text{volume} = \pi r^2 \times h = \pi r^2 h \qquad (6.3)$$

The surface area of a cylinder is equal to the sum of the two end faces (circles), and the curved surface. The curved surface is equivalent to the area of a rectangle equal to the circumference of the end face multiplied by the height of the cylinder.
   Therefore, the total surface area of a cylinder

$$= 2\pi r^2 + 2\pi r h = 2\pi r(r + h) \qquad (6.4)$$

3.1   *Problem 2 – Filling a swimming pool*

A domestic swimming pool 4.5 m in diameter is to be filled with water to a depth of 1.75 m. Calculate how long it will take if water can be transferred at the rate of 0.12 litres/sec (1000 litres = 1 m³).

   The volume of the swimming pool =

$$\pi \times \left(\frac{4.5}{2}\right)^2 \times 1.75 \text{ m}^3$$

or
$$1000 \times \pi \times \left(\frac{4.5}{2}\right)^2 \times 1.75 \text{ litres}$$

The time taken is therefore,

$$1000 \times \pi \times \frac{4.5^2}{4} \times \frac{1.75}{0.12} \times \frac{1}{3600} \text{ hours}$$

The keying sequence is shown in Table A9.

It takes 64 hours 26 minutes to fill the pool.

4 The prism

Any solid which has a constant cross-sectional area parallel to a pair of end faces is called a prism. Thus, the cuboid and cylinder previously discussed are prisms with square or rectangular, or circular cross-sections. Examples of volumes of other constant cross-sectional areas are given below.

4.1 *Problem 3 – Finding the volume of a warehouse*

A temporary warehouse has been constructed with a vertical cross-section in the shape of a major segment of a circle of radius 9 m. The width of the warehouse at the base is 13 m, and it is 52 m long.

Find the internal volume of the warehouse. Assume the material of which the warehouse is made has a negligible thickness.

The cross-sectional area $=$

$$\frac{r^2}{2}(2\pi - \theta + \sin \theta)$$

(see section 9.1, chapter 5)

where r is the radius of the circle and θ is the angle subtended at the centre by a width (w) of the warehouse ($w = BA$ in Figure 5.6, chapter 5).

$$\theta = 2 \sin^{-1} x \times \frac{\pi}{180} \text{ radians} \qquad (6.5)$$

where $\qquad x = \frac{\theta}{2}$ and $\sin x = \frac{\frac{1}{2}w}{r} = \frac{13}{2 \times 9}$

Therefore, the volume of the warehouse =

$$52 \times \frac{81}{2} \left(2\pi - \left(2 \sin^{-1} \left(\frac{13}{18} \right) \times \frac{\pi}{180} \right) + \sin \left(2 \sin^{-1} \left(\frac{13}{18} \right) \right) \right)$$

The keying sequence is shown in Table A10.

The internal volume of the warehouse is 7729 m³.

4.2   Problem 4 – *Finding the volume of a bar*

Find the volume (v) of a bar of length (l) 2.5 m which has a regular hexagonal cross-section of side (d) 12.6 cm, given that

$$v = \tfrac{1}{2}\sqrt{27} \, . \, d^2 l \tag{6.6}$$

The answer is given in Appendix B.

5   Solids of non-uniform height

A solid of uniform cross-section may have a piece sliced off so that its height is non-uniform as in Figure 6.1. Here the solid has a rectangular base $ABCD$, which forms a uniform cross-section up to the plane $EFGH$. The solid part above this has a volume equal to:

$$\tfrac{1}{2}(h_2 - h_1) \times l \times w$$

Therefore, the total volume =

$$(\tfrac{1}{2}(h_2 - h_1) \times l \times w) + (h_1 \times l \times w) = \tfrac{1}{2}(h_1 + h_2) \times l \times w \tag{6.7}$$

that is,

the cross-sectional area of the base × the *average* height.

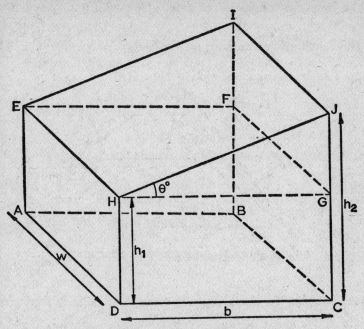

Figure 6.1 A solid of non-uniform height

5.1 *Problem 5 – Using average height*

Find the volume of the solid, as shown in Figure 6.1, which has the following dimensions:

$h_1 = 2.7$ cm, $h_2 = 3.1$ cm, $w = 2.5$ cm and $l = 3.8$ cm.

The answer is given in Appendix B.

6 The pyramid

The pyramid is a solid formed by joining the points on the perimeter of its flat base to a common point (the vertex). Figure 6.2 shows a pyramid with a square base $ABCD$, vertex V.

The volume of a pyramid $= \frac{1}{3}$(area of the base) × perpendicular height   (6.8)

This relationship is true for all pyramids.

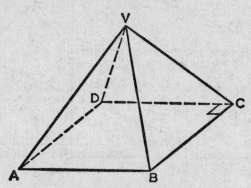

Figure 6.2   A square-based pyramid

6.1   *Problem 6 – Calculating the volume and mass of a wooden stand*

Calculate the volume of a wooden stand which is in the form of a frustum of a pyramid (i.e. the portion remaining when a pyramid is cut by a plane parallel to its base). The two plane faces are squares of side 18 cm and 13 cm respectively, and the height of the frustum is 11 cm. What is the mass of the stand if the density of the wood is 760 kg/m³?

It can be shown that the volume of a frustum of a pyramid is

$$\tfrac{1}{3}h(A_1 + A_2 + \sqrt{A_1 \times A_2}) \qquad (6.9)$$

where h is the height of the frustum and
A_1 and A_2 are the areas of the plane faces of the frustum.

Therefore,

the volume $V = \frac{1}{3} \times 11(18^2 + 13^2 + \sqrt{18^2 \times 13^2})$ cm³

and the mass $= V(\text{m}^3) \times 760$ kg.

The keying sequence is shown in Table A11.

The volume of the stand is 0.00267 m³ and the mass is 2.03 kg.

7 The cone

The volume of a cone can be considered as approximating to a pyramid with many faces and hence is

$\frac{1}{3}$(area of its base) × its perpendicular height,

that is,

$$\tfrac{1}{3}\pi r^2 h \tag{6.10}$$

where r is the radius of the base and h is the perpendicular height. The slant height l is given by,

$$l = \sqrt{r^2 + h^2} \tag{6.11}$$

For a solid cone, the surface area of the cone is a sector of a circle radius l and arc $2\pi r$, plus the area of the base.

That is,

$$\left(\frac{2\pi r}{2\pi l} \times \pi l^2\right) + \pi r^2 = \pi r l + \pi r^2 = \pi r(l + r) \tag{6.12}$$

8 The sphere

The volume of a sphere $= \dfrac{4}{3}\pi r^3$ \hfill (6.13)

where r is the radius of the sphere.

The total curved surface area of a sphere $= 4\pi r^2$ \hfill (6.14)

8.1 *Finding the volume of a portion of a sphere*

Figure 6.3 shows a portion (cap) of a sphere of height h, which has been cut from a sphere of radius r. The radius of the circular base of the cap is R.

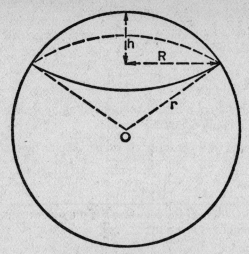

Figure 6.3 Cap of a sphere

It can be shown that the volume of the cap of the sphere in terms of the radius of its base and its height $h =$

$$\frac{1}{6}\pi h(3R^2 + h^2) \tag{6.15}$$

8.2 *Problem 7 – Calculating the mass and weight of a number of locking pins*

Round-headed, tapered metal locking pins have been manufactured to the dimensions shown in Figure 6.4. If the density of the metal used is 8900 kg/m³, calculate how many pins there are to the kilogram.

The head is a cap of a sphere and its volume (V_1) can be found using (6.15), where,

$$h = 5.5 \text{ mm}$$

and $R = 3.5 \text{ mm}$

i.e. $V_1 = \dfrac{1}{6} \times \pi \times 5.5((3 \times 3.5^2) + 5.5)$

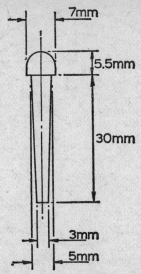

Figure 6.4 Locking pin

The body of the pin forms a frustum of a cone, and its volume (V_2) can be found using (6.16) below.

$$V_2 = \tfrac{1}{3}\pi H(r_1{}^2 + r_1 r_2 + r_2{}^2) \tag{6.16}$$

where,

the height of the frustum $H = 30$ mm

and

the radii of the plane faces r_1, $r_2 = 1.5$ mm and 2.5 mm respectively.

Therefore, the volume of the pin is

$$\tfrac{1}{3}\pi(30(1.5^2 + (1.5 \times 2.5) + 2.5^2) + \frac{5.5}{2}((3 \times 3.5^2) + 5.5))$$

Multiply by 8900×10^{-9} to obtain the mass of each pin in kg. The number of pins per kilogram is the reciprocal of this result.

The keying sequence is shown in Table A12.

There are 222 pins to the kilogram.

Part II
Applications

7 Statistics

1 Introduction

If you intend to use your calculator extensively for statistical calculations, it may be worthwhile obtaining a calculator with specialised statistical keys. The general sequence of sophistication offered is:

- (a) normal scientific plus mean and standard deviation functions,
- (b) as (a) plus factorial and $n!/r!(n-r)!$ functions,
- (c) as (b) plus trend and regression.

Although these calculators appear to be the natural choice for statistical calculations, the additional keys may prove to be more limited than the prospective buyer anticipates.

Calculators having mean and standard deviation keys usually require that the individual readings are entered; they cannot be entered in grouped form. This means that with standard deviation calculations based upon frequency tables, it is often just as convenient, less prone to error, and quicker to carry out the calculation in the conventional manner for grouped data.

Some calculators having a standard deviation function also have keys to access the intermediate results (i.e. Σx, Σx^2, etc.). This may be a useful additional feature, particularly if calculations associated with linear regression are likely to be required. A point to note with calculators having these specialised keys (standard deviation, Σx, Σx^2, etc.) is that some calculators allow the values to be accessed (recalled) only once and subsequent use of the key leads to error conditions. With other calculators the values can be regarded as being in special-purpose memories to be recalled any time.

Calculators having factorial keys may be extremely useful if many calculations involving these keys are intended. In the field of statistics, they may be relevant to calculations based upon the multiplication and addition laws of probability, which develop into calculations involving permutations and combinations. A particular application would be where calculations involving the Binomial Distribution are required.

The third type of specialised calculator mentioned above, with trend and regression keys, needs careful assessment. In science, fitting a straight line to a series of points is often done and such a feature would be a useful addition to the calculator. In business applications, more caution is needed as there are many other approaches to forecasting.

Techniques such as EWMA (exponentially weighted moving averages) are much more commonly used nowadays and can be conveniently evaluated on a standard calculator (see p. 101).

The calculators having a few statistical keys are therefore still very limited in statistical capability. For the majority of statistical calculations, a conventional scientific calculator will be almost as good. Although it is difficult to generalise, perhaps statistical calculations require more carrying forward of parameters than other applications so that, other facilities being equal, the more memories the better, i.e. a standard scientific calculator having two memories may prove more useful than one having a single memory plus standard deviation function.

To avoid mistakes in statistical calculations, it is important to be neat and systematic in the calculations. Manual statistical methods are often based upon a tabular presentation and development of the data. Until you are proficient in the use of the calculator, and the means of carrying through intermediate results via the memory, it is safer to preserve the tabular approach and note down intermediate results in the conventional manner.

The field of statistics is very large and it is necessary to cover in this chapter the straightforward applications of formulae using a calculator. This chapter concentrates on those areas where the use of a calculator may lead to a slightly different method of calculation, allows short cuts, or suggests a different approach.

2 Standard deviation of grouped data

The basic formula for calculating a standard deviation is often expressed in one of several suitable forms for calculation such as:

$$S.D. = \sqrt{\frac{1}{n}\left[\Sigma fx^2 - \frac{(\Sigma fx)^2}{n}\right]} \qquad (7.1)$$

where
$$x = \text{class mid-points,}$$
$$f = \text{associated frequencies}$$
$$n = \Sigma f$$

Although this formula can be evaluated on a calculator without any operational problems, it is useful to know that the data can often be coded to give simpler arithmetic calculations. In the past, the purpose of coding has been to simplify the arithmetic so that mistakes are avoided. With a calculator the need to deal with simple numbers is not important from an arithmetic point of view, but it does affect the number of key depressions required. Therefore, there is still a need to consider methods that lead to simpler keying, as this will ensure the correct values are entered and the correct sequence is being followed.

The principles of coding data are based upon two considerations:

(a) A constant quantity can be added or subtracted to each result. This will not affect the standard deviation but merely shift the mean by a known amount.

(b) All the measurements may be scaled up or down by a constant factor. This will similarly scale the standard deviation in a corresponding manner.

To simplify the calculations using the above principles, an assumed mean can be established at the midpoint of the first class interval and a scaling factor equivalent to the class interval used.

The keying sequence for Σfx and Σfx^2 now involves simpler numbers arranged in such a way that it is less likely that a row will be omitted.

It can be shown that:

$$\Sigma fx = 0 \,.\, f_1 + 1 \,.\, f_2 + 2 \,.\, f_3 + \ldots (n-1)f_n \qquad (7.2)$$

and

$$\Sigma fx^2 = 0 \,.\, f_1 + 1^2 \,.\, f_2 + 2^2 \,.\, f_3 + \ldots (n-1)^2 f_n \qquad (7.3)$$

With calculators having two memories, these expressions are particularly appropriate as they can be evaluated concurrently as illustrated by the following problem.

2.1   Problem 1 – Calculation of mean and standard deviation

Table 7.1   Coding data

Original Data			Coded Data	
Class	Midpoint	Frequency	x	f
100—	110	25	0	25
120—	130	40	1	40
140—	150	63	2	63
160—	170	37	3	37
180—	190	20	4	20

assumed mean = 110, scaling factor = 20

185 readings fall between the values 100 and 200 as shown in Table 7.1; the data have been coded as shown in the last two columns of the table. Because the interval between class midpoints is 20, this has been chosen as the scaling factor. Thus each subsequent class interval is one scaled unit higher. For convenience the midpoint of the first class interval has been chosen as the assumed mean, i.e. the constant 110 has been subtracted from each class midpoint prior to scaling. For example, the class interval 140–160 (midpoint 150) is coded by subtracting 110 and then dividing by the scaling factor, i.e.

$$\frac{150 - 110}{20} = \frac{40}{20} = 2$$

It is required to calculate the mean and standard deviation of the coded data and decode in terms of the original data. The straightforward approach would be to evaluate Σfx and note the answer, then evaluate Σfx^2 and note this, then to substitute these noted values and n in (7.1) to find the standard deviation. However, as mentioned above, with two memories the keying sequence given in Table 7.2 can be used to take advantage of the relationship between (7.2) and (7.3). Table 7.2 illustrated the calculation of fx and fx^2 and the concurrent summation of each in two memories.

Table 7.2 Use of two memories

Keying	Display	Comments
1		first row (Table 7.1) = zero
×		
40		
=	40.	fx, second row
M1+		add to memory 1
×		
1		x
=	40.	fx^2
M2+		add to memory 2
2		x, third row
×		
63		
=	126.	fx, third row
M1+		add to memory 1
×		
2		x, third row
=	252.	fx^2
M2+		add to memory 2
etc.		

Whichever approach is adopted, the coded data in Table 7.1 lead to:

$$\Sigma fx = 357$$
$$\Sigma fx^2 = 945$$

Thus the coded mean $= \dfrac{\Sigma fx}{n} = \dfrac{357}{185} = 1.93$

and the coded standard deviation $= 1.1765$ (see Table A13).

Note: that in evaluating $\Sigma fx^2 - (\Sigma fx)^2/n$, it is more convenient to key in and square Σfx first.

To decode the results, it is necessary to take into account the scaling factor and assumed mean. The actual mean is apparently 1.93 scaled units above the assumed mean. Therefore,

$$\text{actual mean} = 110 + (1.93 \times 20) = 148.6$$

The standard deviation can be similarly calculated, i.e.

$$\text{actual standard deviation} = 1.1765 \times 20 = 23.53.$$

2.2 Problem 2 – Coding exercise

Calculate the mean and standard deviation of the data given in Table 7.3.

Table 7.3 Coding exercise

Class	Frequency
0—	25
10—	50
20—	45
30—	20
40—	10

The table needs to be expanded to show the class midpoints and coded x and f values in a similar manner to Table 7.1.

The intermediate and final results are given in Appendix B.

3 The area under the normal distribution curve

Usually, even when a standard deviation has been evaluated on a calculator the interpretation is by means of a normal distribution table. However, it is possible to calculate the area under the curve in a normal distribution, although the procedure is somewhat involved.

The formula given evaluates the upper tail of the distribution where the deviation is $\geqslant 0$ (see Figure 7.1).

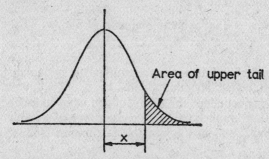

Figure 7.1 Calculating the area of the upper tail of the Normal distribution

The probability of x or more can be represented by the following approximation:

$$P(\geqslant x) = ((((t \cdot b_5 + b_4)t + b_3)t + b_2)t + b_1)t \cdot h \cdot k \quad (7.4)$$

where,

$$t = (1 + 0.2316419x)^{-1}$$
$$b_1 = .31928153$$
$$b_2 = -.356563782$$
$$b_3 = 1.781477987$$
$$b_4 = -1.821255978$$
$$b_5 = 1.330274429$$
$$h = e^{-x^2/2}$$
$$k = \frac{1}{\sqrt{2\pi}}$$

The formula has been presented in nested parenthical form to suit the keying requirements of a calculator having a single memory.

Given a value for x, t can be evaluated and retained in the memory allowing a chain calculation to be used to evaluate the nested parentheses. This interim result is then stored while h is evaluated and used to modify the interim result. The second interim result is retained in the memory while k is evaluated.

3.1 *Problem 3 – Calculating the area in the tail of the normal distribution*

It is required to find the area in one tail of the normal distribution when the standard deviation (x) is 2.

The keying sequence for this problem is given in Table A14. The calculated result is 0.0227 and compares favourably with the values to be found in the normal distribution table (Appendix D).

Note: For simplicity, not all the significant figures have been entered in the keying table.

3.2 *Finding the deviation given the area*

The inverse of this situation is when it is required to find x given the area in one tail of the normal distribution. A formula for estimating x from a given probability (p) is shown below:

$$x = t - \frac{(c_2 t + c_1)t + c_0}{((d_3 t + d_2)t + d_1)t + 1} \qquad (7.5)$$

where

$$t = \sqrt{\ln\left(\frac{1}{p^2}\right)}$$

$$
\begin{array}{ll}
c_0 = 2.515517 & d_1 = 1.432788 \\
c_1 = 0.802853 & d_2 = 0.189269 \\
c_2 = 0.010328 & d_3 = 0.001308
\end{array}
$$

A general keying sequence for (7.5) is given in Table 7.4.

Table 7.4 Keying sequence for deviation given area

Keying	Display	Comments
p		area as a probability
x^2		
1		
$\dfrac{1}{x}$		
ln		
$=$		$ln\left(\dfrac{1}{p^2}\right)$
$\sqrt{x}$		t
M+		store t
×		
d_3		0.001308
+		
d_2		0.189269
×		
MR		t
+		
d_1		1.432788
×		
MR		t
+		
1		
$=$		note down, call U
c_2		0.010328
×		
MR		t
+		
c_1		0.802853
×		
MR		t
+		
c_0		2.515517
÷		
U		previously noted result
M−		
MR		answer

4   Generating random numbers

A simple pseudo-random number generator has been reported
in a Hewlett-Packard applications manual, using the relationship:

$$r = \text{fractional part of } (\pi + r_{-1})^5 \qquad (7.6)$$

where the initial value of r is specified as a 'seed' such that,

$$0 \leqslant r \leqslant 1$$

and r_{-1} represents the previous value of r.

Using 0.1234567 as a seed, the keying sequence is given in
Table 7.5. The first number generated is .06 or 06 as a two-digit
number in this table. However, as this method is sensitive to the
values of the last few significant figures displayed, the actual
pseudo-random number generated will vary according to the
calculator used.

Table 7.5   Generation of pseudo-random numbers

Keying	Display	Comments
.12345		seed, five decimal places entered
+		
π		
=	3.2650	
ln	1.1832	
×		
5		
=	5.9163	
e^x	371.06	integer + random decimal
−		
371		
=	0.06013	random number
etc.		repeat with random number as new 'seed'

5   Regression and correlation

Books on statistics present many forms of the relationships for
establishing the equation of a straight line (linear regression).

Using the formulae presented here, a systematic way of establishing the equation of the line, the coefficient of correlation and the residual standard deviation is obtained.

The basis is to calculate three intermediate sums,

$$S_{xy}, S_{xx}, S_{yy}, \text{ where}$$

$$S_{xy} = \Sigma xy - \frac{\Sigma x \Sigma y}{n} \tag{7.7}$$

$$S_{xx} = \Sigma x^2 - \frac{(\Sigma x)^2}{n} \tag{7.8}$$

$$S_{yy} = \Sigma y^2 - \frac{(\Sigma y)^2}{n} \tag{7.9}$$

the data consisting of n pairs of x, y values.

From these the coefficient of correlation, r, can be easily found,

i.e.
$$r = \frac{S_{xy}}{\sqrt{S_{xx}S_{yy}}} \tag{7.10}$$

If the data appear to be acceptably correlated, the equation can be established from,

$$\text{slope} = m = \frac{S_{xy}}{S_{xx}} \tag{7.11}$$

and the value of x at $y = 0$,

i.e.
$$c = \bar{y} - m\bar{x} \tag{7.12}$$

where $\bar{y}$ and $\bar{x}$ are the means of the original data.

From these calculations, an equation of the following form can be developed,

$$\hat{y} = mx + c$$

The accuracy of estimates of y ($\hat{y}$) based upon this equation can be assessed by calculating next the residual standard deviation.

$$\text{Residual S.D.} = \sqrt{\frac{S_{yy} - mS_{xy}}{n - 2}} \tag{7.13}$$

There are no inherent problems in using a calculator for the abov
and the systematic approach suggested should ensure that keyi
is accurate.

5.1   *Problem 4 – Regression calculation*

Table 7.6   Development of data for regression analysis

x	y	x^2	y^2	xy
112	70	12 544	4 900	7 840
116	90	13 456	8 100	10 440
104	50	10 816	2 500	5 200
101	40	10 201	1 600	4 040
433	250	47 017	17 100	27 520

The systematic development of data is shown in Table 7.6. Fro
this table, the following evaluations can be made.

$$\bar{x} = \frac{433}{4} = 108.25$$

$$\bar{y} = \frac{250}{4} = 62.5$$

$$S_{xy} = 27\,520 - \frac{(433)(250)}{4} = 457.5$$

$$S_{xx} = 47\,017 - \frac{(433)^2}{4} = 144.75$$

$$S_{yy} = 17\,100 - \frac{(250)^2}{4} = 1475$$

5.2   *Problem 5 – Completion of calculations*

Use the values of S_{xy}, S_{xx}, S_{yy}, found in Problem 4, to calculat
r (7.10), m (7.11) and c (7.12).

What is the residual standard deviation?

The answers are given in Appendix B.

6 Cumulative sum techniques and detecting trends

Cumulative sum techniques provide a way of detecting a change in an average value very quickly.

The basic procedure is to subtract some constant quantity from a series of figures and to calculate the resulting cumulative differences, (sometimes referred to as the 'Cusum'). This is particularly easy to do on a calculator and therefore provides a practical alternative to other trend-spotting techniques.

Providing the average of the series of figures is close to the constant quantity that is being subtracted, the differences will be positive or negative causing the Cusum to fluctuate slightly about some particular value. If there is a change in the average of the original data, the differences will be predominately positive *or* negative, resulting in the Cusum adopting a marked upward *or* downward slope respectively. The sensitivity of the Cusum in detecting small changes in the average value is best shown by an example.

6.1 Problem 6 – Detection of change in the average

The example comprises two sets of data (see Table 7.7). The first set is 20 figures varying at random about a mean of 50, and the second set is a further 20 random figures about a mean of 70.

If the two sets of figures are plotted in the normal manner, they show no discernible change. However, when they are converted into a Cusum, the change in average value is quickly detected as shown in Figure 7.2. The keying is very simple and an example of the keying sequence for the first few rows of Table 7.7 is shown in Table 7.8.

Table 7.7 Cusum applied to a series

Value	Difference (from 50)	Cusum
30	−20	−20
49	−1	−21
58	+8	−13
50	0	−13
106	+56	+43
38	−12	+31
5	−45	−14
98	+48	+34
70	+20	+54
11	−39	+15
60	+10	+25
1	−49	−24
48	−2	−26
25	−25	−51
63	+13	−38
87	+37	−1
37	−13	−14
82	+32	+18
61	+11	+29
39	−11	+18
75	+25	+43
76	+26	+69
60	+10	+79
103	+53	+132
21	−29	+103
78	+28	+131
32	−18	+113
93	+43	+156
47	−3	+153
92	+42	+195
117	+67	+262
90	+40	+302
58	+8	+310
43	−7	+303
18	−32	+271
109	+59	+330
50	0	+330
92	+42	+372
104	+54	+426
116	+66	+492

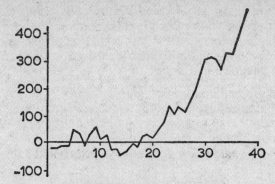

Figure 7.2 Plot of a Cusum

7 Exponential smoothing

Exponential smoothing is commonly used to smooth time series. One advantage of exponential smoothing is that greater weight is attached to the latest data, while older data are correspondingly discounted. The nature of the weighting used means that the technique is also often called EWMA (exponentially weighted

Table 7.8 Calculation of Cusum

Keying	Display	Comments
50		estimated 'constant' difference
M+		store as constant
30		
−		
MR	50.	
=	−20.	first Cusum
+		
49		
−	29.	
MR		
=	−21.	second Cusum
+		
58		
−	37.	
MR		
=	−13.	third Cusum
etc.		

moving averages). The method illustrated is only the simplest of many variations developed in this area.

A further advantage of EWMA is that successive calculations only require one value to be carried forward. This can be seen from the relationship used,

New smoothed average =
   a (latest result) + (1 − a) (old smoothed average)   (7.14)

where,

a = smoothing constant, between 0 and 1.0.

7.1   *Problem 7 – Forecasting the next value in a time series*

Table 7.9   Past sales data

Year	Sales
73	30
74	28
75	34
76	32
77	36
78	?

It is required to forecast the next value in the time series given in Table 7.9. Initially the old smoothed average is taken to equal the first value given. This always leads to the forecast for the second period being the same as the first value. A typical value for the smoothing constant is 0.2.

From the keying sequence given in Table A15, the forecast for 78 is seen to be 31.83.

7.2   *Problem 8 – Updating the forecast*

If the actual value for 1978 in Table 7.9 turns out to be 33, what would be the forecast for 1979?

The answer is given in Appendix B.

8 Finance and Accountancy

1 Introduction

This chapter concentrates on calculations involving money. Section 2 is an extension of simple calculations typically associated with retailing and selling. Section 3 is related to the 'purer' financial calculations used in investment. In this section, because of the more complicated formulae, it is assumed that the reader is familiar with the more basic operations of the calculator. The remaining sections, 4 and 5, are illustrative of the types of calculations that would interest an accountant.

It should be emphasised that the correct use of the formulae quoted and precise definitions of accountancy terms used are outside the scope of this book.

2 Retailing and selling

Calculations in this area are largely pro-rata and/or percentage calculations. The problems presented here are intended to give additional practice in this type of work.

2.1 *Problem 1 – Cash discounts*

GT Motor Accessories ordered five dozen wheel covers at £2.15 each, two dozen key fobs at £0.40 each and three dozen driving gloves at £3.10 each. They will qualify for a 5% discount. For how much will GT Motor Accessories be invoiced?

The calculation is essentially chain arithmetic, with the intermediate costs being cumulated in the memory prior to calculating the final discount. The keying sequence is given in Table A16 and gives an invoiced total of £237.69.

There are two points to note in calculating this answer.

(a) As all the quantities are dozens, the individual entries would have been in dozens, i.e. 5 × 2.15 etc. The final cumulative total (or indeed the final invoice value) could then be scaled up once by a factor of 12.

(b) In Table A16, the discount is calculated and then 'subtracted'. Alternatively, the 'complement' of the discount could be used to scale down the total. That is, 5% discount means goods are purchased at 95% of full price, i.e. 250.2 × .95. This cuts out the last three keying operations.

2.2   *Problem 2 – Invoice calculation*

A customer buys six metres of timber at 75p a metre, one tin of adhesive at 89p and two sq. metres of chipboard at £1.40 a sq. metre, plus 50p cutting charge. How much should he pay?

The answer is given in Appendix B.

2.3   *Problem 3 – Price adjustment*

A confectioner buys Easter eggs wholesale at 70p. They have a mark-up of 60%. After Easter the eggs are reduced by 25%. What is the sale price of the eggs?

It is wrong to simply take the final mark-up as 60% − 25%, i.e. 35%. The correct keying sequence is shown in Table A17. From this it is seen that the sale price will be 84p. Table A17 has been written out to show the obvious (but long) way of calculating the answer. By comparison, the keying sequence in Table A18 is much shorter. This illustrates how a moment spent preplanning the approach can save keystrokes and thereby time and the likelihood of keying errors.

2.4   *Problem 4 – Calculating commission*

An agent gets 25p in the £1.00 commission and £5.00 for every new client introduced. In the last month she sold £350.00 of goods and introduced three new clients. How much will she earn?

The answer is given in Appendix B.

3 Investment appraisal calculations

There are many methods of assessing the worthiness of a proposed investment. The traditional payback period method is attractive because of the simplicity of calculations. As the possibility of more sophisticated analysis has presented itself with computers and now pocket calculators, a wider range of alternative methods becomes practical. Even so, no one method overcomes all the shortcomings of the others and so it is still up to the analyst to choose the most appropriate. It is not the purpose of this book to assess the strengths and weaknesses of the many methods from a financial point of view.

In practice, many of the methods are concerned with taking into account the cash flow over time and many of the appraisal methods are just variations on a Discounted Cash Flow concept.

3.1 *Basic financial formulae*

The bulk of financial calculations involve making use of some of the following five parameters:

Present value: This represents the value of future cash flows discounted back to the present. For practical purposes, it can also be equated to the size of a loan being negotiated now (e.g. a mortgage).

Future value: This represents the future value (as a lump sum) at a specified point in time of previous cash flows. For example, a Building Society investment plan will result in a specified amount of capital being accumulated after say five years, from the regular payments.

Uniform payments/ *re-payments:*	This represents the regular cash flow required to pay off (repay) a loan, or the specified future lump sum.
Number of periods:	This represents the number of uniform periods involved in the calculation.
Interest rate:	This represents the rate at which the capital involved is 'growing' or the rate at which a future value is 'discounted' back to the present. As this is a rate, it must be in the same time units as the periods (e.g. note that 2% per month is not 24% per annum, see also p. 108).

In practice, problems involving combinations of these five parameters lead to a dozen relationships being commonly used. The twelve formulae are presented in Table 8.1 together with a 'look up' table to assist in the selection of the appropriate relationship. The bulk of the examples in this chapter make use of, or are related to, these formulae.

3.2 *Calculating mortgage repayments*

The factors involved in calculating repayments are:

The size of the loan, i.e. principal (P)

The duration of the loan (n years)

The interest rate (i% per annum)

The appropriate formula to use is no. 11 in Table 8.1, although a common alternative form of presentation is:

$$\frac{Pi(1 + i)^n}{(1 + i)^n - 1}$$

This formula is not so convenient to evaluate on a calculator as it stands, but it can be rearranged by dividing throughout by $(1 + i)^n$. This then leads to the expression,

$$\frac{Pi}{1 - \dfrac{1}{(1 + i)^n}}$$

Table 8.1 Look-up table of financial formulae

This table shows the appropriate formulae to use given three input parameters and a required output parameter.

GIVEN		1	2	3	4	5	6	7	8	9	10	11	12
no. of uniform periods	(n)				X	X	X	X	X	X	X	X	X
interest rate per period	(i)	X	X	X				X	X	X	X	X	X
present value	(P)	X	X		X	X				X		X	
future value	(F)	X		X	X		X	X					X
uniform payments	(R)		X	X		X	X		X		X		

REQUIRED	
n	1 2 3
i	4 5 6
P	7 8
F	9 10
R	11 12

1. $n = \dfrac{\log \dfrac{F}{P}}{\log (1 + i)}$

2. $n = \dfrac{\log \dfrac{1}{1 - iP/R}}{\log (1 + i)}$

3. $n = \dfrac{\log \left(1 + \dfrac{iF}{R}\right)}{\log (1 + i)}$

4. $i = \sqrt[n]{\dfrac{F}{P}} - 1$

5. $i = \dfrac{R}{P}(1 - (1 + i)^{-n})$

6. $i = \dfrac{R}{F}([1 + i]^n - 1)$

7. $P = \dfrac{F}{(1 + i)^n}$

8. $P = R\,\dfrac{1 - (1 + i)^{-n}}{i}$

9. $F = P(1 + i)^n$

10. $F = R\,\dfrac{(1 + i)^n - 1}{i}$

11. $R = P\,\dfrac{i}{1 - (1 + i)^{-n}}$

12. $R = F\,\dfrac{i}{(1 + i)^n - 1}$

Notes:

(a) where logarithms are required either $\log_{10}$ (and 10^x) or ln (and e^x) may be used.

(b) 5 and 6 are solved iteratively.

The general sequence of calculations on any calculator is then:

Evaluate	*Comment*
$1 + i$	
$(1 + i)^n$	raise to power n
$\dfrac{1}{(1 + i)^n}$	take reciprocal
$1 - \dfrac{1}{(1 + i)^n}$	bottom line of formula
$\dfrac{1}{1 - \dfrac{1}{(1 + i)^n}}$	take reciprocal
$\dfrac{Pi}{1 - \dfrac{1}{(1 + i)^n}}$	complete by multiplying by P, then i.

In evaluating this expression, it is important that the working is in the correct units. If n is in years, then i is the % interest per annum and the answer is the *annual* repayments. In practice, most people wish to calculate the *monthly* repayment. It is not one twelfth of the annual repayment. Rather the expression has to be evaluated for n in months and hence i is the % interest per month.

3.2.1   *Problem 5 – Calculating monthly repayments*

What monthly repayments are required for a £10 000 loan over twenty years when the interest rate is 1 % per month?

As n needs to be expressed in months, i.e. 240, there is the need to evaluate $(1 + 0.01)^{240}$ on substitution in the formula. Providing $(1 + 0.01)^{240}$ can be evaluated, the remainder of the calculation is straightforward. Depending upon the calculator's facilities, y^x can be used (see Table A19), or the log and 10^x keys can be used, or ln and e^x keys (see Table A20) or finally the x^2 key may be used. The use of these facilities is described in chapter 4.

To use the x^2 key, it is necessary to find the equivalent of x^{240} in factors of x^2. This idea can be developed algebraically, as follows:

$$x^{240} = x^{256-16}$$

$$= \frac{x^{256}}{x^{16}}$$

$$= \frac{((((((x^2)^2)^2)^2)^2)^2)^2}{(((x^2)^2)^2)^2}$$

As the bottom line will be evaluated in the course of working out the top line, the intermediate result can be noted or put into memory. The keying sequence required is given in Table A21.

Whatever method is used to evaluate $(1 + 0.01)^{240}$ the final part of the calculation is shown in Table A22. From this it can be seen that the monthly repayments are £110.11.

3.3 Calculating length of loan

Transposition of the previous formula allows the complementary question to be answered, namely, for a given monthly repayment how long would it take to pay off the loan?

The usual transposition is:

$$(1 + i)^n = \frac{R}{R - Pi}$$

where R = uniform repayments.

This relationship can be rewritten so that R only appears once, thereby making it more suitable for chain calculation, by dividing the right hand side throughout by R,

i.e.
$$(1 + i)^n = \frac{1}{1 - \dfrac{Pi}{R}}$$

As before if R is a monthly repayment then n and i should be expressed in months. The calculation is best tackled if possible by the use of log functions, so that:

$$n = \frac{\log\left(\dfrac{1}{1 - \dfrac{Pi}{R}}\right)}{\log(1 + i)} \quad \begin{array}{l}\text{(see formula 2,}\\ \text{Table 8.1)}\end{array}$$

If the calculator has log keys, this presents no problems. If not, it will usually be necessary to complete the calculation using log tables. A more awkward way is to evaluate the right hand side:

$$\frac{1}{1 - \dfrac{Pi}{R}} \quad \text{and note answer}$$

Then enter $(1 + i)$ as a constant and count the number of times it is necessary to multiply it by itself before it reaches the right-hand side value. The problem with this approach is that you are likely to be working in monthly time units in which case a twenty-year mortgage leads to at least 240 key depressions.

3.3.1 *Problem 6*

If the loan is for £10 000 at 12.5% p.a. and it is intended to repay at the rate of £200.00 per month, find the required duration of the loan. The keying sequence for this problem is shown in Table A23 and gives the answer five years eleven months.

3.4 *Making a table of present worth factors*

To convert a future value to a present value, it is necessary to use formula 7, Table 8.1. In practice, this formula can be rewritten:

$$P = S \times \text{a factor}$$

The factor is commonly termed the Present Worth Factor (PWF) and is algebraically =

$$\frac{1}{(1 + i)^n}$$

To compile a table from any starting value of n, the successive value is found by dividing the previous value by $(1 + i)$ held in the memory.

3.4.1 *Problem 7*

Find the present worth factors for years five to eight for a 15%

discount rate. The keying sequence is given in Table A24 and yields the factors: .49718, .43233, .37594, .32690.

3.4.2 *Problem 8*

What are the present worth factors for a 20% discount rate for years three to five inclusive? The answers are given in Appendix B.

3.5 *Miscellaneous examples and problems associated with Table 8.1*

3.5.1 *Problem 9*

I wish to find an investment to make £3000 grow to £6000 in seven years. What rate of return must I achieve?
  This can be solved using formula 4, where $P = 3000$, $F = 6000$ and $n = 7$. The keying sequence is given in Table A25 and shows that a 10.4% annual return is required.

3.5.2 *Problem 10*

In the previous problem, I only seem to be able to achieve a 9% growth rate. How much will I actually end up with after seven years? The answer using formula 9 in Table 8.1 is given in Appendix B.

3.5.3 *Problem 11*

Using formula 10, Table 8.1, check that with an interest rate of 6.5%, £500 p.a. will accumulate to approximately £12 000 in fifteen years.

The answer is given in Appendix B.

4 Financial ratios

Another area for the use of a calculator is in the interpretation of balance sheets and statements. The calculations are simple and require little more than four function chain calculations. The ratios calculated below are only indicative of the type of calculations, hence no precise definition of terms is given.

4.1 *Problem 12 – Analysis of accounts*

Table 8.2 Extract from accounts of XYZ Co.

Year	Sales £000 (A)	Profit £000 (B)	Net Assets £000 (C)	Stocks £000 (D)
1976	12 680	2541	6872	4217
1977	15 660	2898	7200	4628

Extracts from the accounts of the XYZ company are shown in Table 8.2. Analyse the 1976 figure to find:

(*a*) Return on investment.
(*b*) % profit.
(*c*) Rate of stock turnover.

The keying sequence for these calculations is shown in Table A26 from which it can be seen that:

$$\text{Return on investment} = 37\%$$
$$\% \text{ profit} = 20\%$$
$$\text{Rate of stock turnover} = 3.0$$

To save keying, the final stage of multiplying by 100 to obtain a % has been omitted.

4.2 *Problem 13 – Updating analysis*

Repeat the above analysis for the 1977 figures.

The answers are given in Appendix B.

5 Reducing balance method of depreciation

With the reducing balance method, it is necessary to calculate the factor to apply to the reducing balance so that starting with an initial cost of an asset it will end up at its scrap value after a specified number of years.

The formula is: factor $= \left(\dfrac{S}{C}\right)^{1/n} = \sqrt[n]{\dfrac{S}{C}}$

where,

C is the initial cost,
S is the required final scrap value,
n is the required number of years.

5.1 *Problem 14 – Depreciation factor*

If the initial cost of a fork truck is £6000 and it is required to depreciate the truck down to £250 over ten years, what depreciation factor should be used?

The keying sequence is shown in Table A27 and gives a factor of 0.7277444. If this depreciation factor is held in the memory, then successive depreciated values may readily be obtained by multiplying the latest value by the memory contents. The results are shown in Table 8.3.

Table 8.3 Depreciated values

		£	
Initial cost		6000	
Year ending	1	4366.47	(i.e. 6000 × .7277444)
	2	3177.67	
	3	2312.53	
	4	1682.93	
	5	1224.74	
	6	891.30	
	7	648.64	
	8	472.04	
	9	343.53	
	10	250.00	

5.2   *Problem 15 – Depreciation of a car*

It is required to depreciate a £5000 car to £2000 over four years. What factor should be used?

The answer is given in Appendix B.

9 Management

1 Introduction

The use of calculators in management implies a quantitative approach. Some of the areas of interest have already been covered on the chapters on Statistics and Finance and Accountancy. The problems covered in this chapter are more related to the operational side and the reader is referred to *Teach Yourself Operational Research* (Makower and Williamson) for amplification of some of the formulae used.

2 Break-even analysis

As the financial aspects of break-even analysis are a means to an end rather than an end in themselves, this section is included under Management not Finance.

Break-even analysis relates the fixed and variable costs to the volume of production (or sales). Often this type of calculation is required when a change in process or procedure is being investigated.

A change in the operations is often accompanied by a change in level of production. There are thus five cost points of interest as shown in Figure 9.1. These are the cost comparisons of two processes A and B, at present production levels (a_1 and b_1), the cost comparison at some future proposed production level (a_2 and b_2) and the position of the break-even point (e).

2.1 *Problem 1 – Selecting a production process*

The fixed costs associated with producing metal garden sprinklers are £3000, while the variable cost is 50p each. Plastic sprinklers

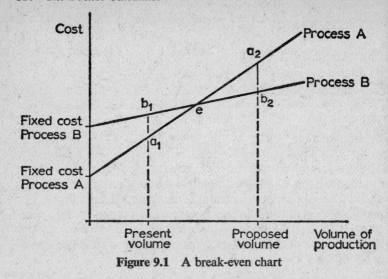

Figure 9.1 A break-even chart

can be produced with a fixed cost (including moulds) of £7500 and a variable cost of 20p. What is the break-even point volume of production?

$$\text{Break-even volume} = \frac{\text{Fixed cost } B - \text{Fixed cost } A}{\text{Variable cost } A - \text{Variable cost } B} \quad (9.1)$$

The keying sequence for this calculation is shown in Table A28 and gives a volume of 15 000 sprinklers.

2.2 *Problem 2 – Profit or loss?*

The manufacturer in the above problem has been advised that at £2.00 each the total sales will be 10 000 while at £1.00 each, the market will increase to 25 000. Which price leads to the larger gross 'profit' and which process should be used?

The approach is to calculate the total cost for the cheaper process at an output volume of 10 000 (i.e. process A) and from this establish the profit. The profit is then compared with a similar calculation involving process B at the output volume of 25 000.

The answer is given in Appendix B.

3 Stock control

3.1 *Problem 3 – Order quantity*

The manufacturer of sprinklers uses 10 000 plated nozzles a year. They cost him 18p each. It costs £5.00 to raise the paperwork to place an order. The cost of tying up capital in stock and holding stock is 15% of the unit price. What is the economic quantity to order?

$$\text{Economic order quantity} = \sqrt{\frac{2cd}{ip}} \qquad (9.2)$$

where,
c = cost of placing order
d = annual usage
i = % holding cost
p = unit price

The keying sequence is shown in Table A29 and gives an order quantity of 1925, say 2000.

3.2 *Problem 4*

The manufacturer also buys in the cast base at 30p each. What is the economic order quantity for these?

The answer is given in Appendix B.

3.3 *Problem 5 – Standby stock*

A processing plant depends upon the continued working of a motor unit. As it would take six months to obtain another motor unit from the supplier, a standby stock is held. The cost of holding additional motor units is £260 per unit and the cost of not having a motor unit when required is £580. The probable demand for spares during the six month reordering period is as shown in Table 9.1. How many additional motor units should be held?

Table 9.1   Probable demand for motor unit

Demand (also standby stock)	Probability of this demand *or more*
0	1.00
1	.80
2	.75
3	.50
4	.34
5	.20
6	.10
7	.05
8	.01

The probability of a stock-out $\leqslant \dfrac{h}{h+s}$ (9.3)

where h = cost of holding extra items when not required
s = cost of not having item when required.

The keying sequence in Table A30 shows that the probability of a stockout should not exceed 0.309. Inspection of Table 9.1 shows that the minimum standby stock (or demand) that can be accepted is five units.

3.4   *Problem 6 – Modified process*

The process is modified so that the cost of not having an operative motor is reduced to £370. What is the revised stock level of standby motors?

The answer is given in Appendix B.

4   Activity sampling

4.1   *Problem 7 – Machine utilisation*

2000 observations were carried out in a factory to determine the causes of loss of production. The results are given in Table 9.2.

What is the utilisation of the machine and how accurate is this answer?

Table 9.2   Activity sampling study

Activity	Number of occurrences
Machine operating	1253
Setting-up	356
Waiting for material	235
Waiting for instructions	120
Breakdown	36
Total	**2000**

The machine utilisation can be found simply by expressing the proportion of time operating to the total as a percentage. The accuracy of this result is given by

$$\text{Limits} = \pm 2 \sqrt{\frac{P(100 - P)}{N}} \qquad (9.4)$$

where $P = \%$ occurrence

N = total number of observations

and 2 represents two standard deviations to give approximately 95% confidence.

The keying sequence is shown in Table A31 and gives an answer of $62.65\% \pm 2.16\%$.

4.2   *Problem 8 – Breakdown time*

What is the accuracy of the % occurrence of machine breakdowns? If management want the answer to be within $\pm 0.5\%$, how many *more* observations are required? (Transpose the formula to express N in terms of P and L and use the value of L just calculated.)

The answer is given in Appendix B.

5   Quality levels

5.1   *Problem 9 – Quality check*

A car dealer expects on average to find fifteen minor items to rectify on each new car delivered. Within what limits are the actual number of rectifications likely to vary if there is no change in quality?

$$\text{Warning limits} = \pm 1.96\sqrt{\text{mean}} \tag{9.5}$$

$$\text{Action limits} = \pm 2.05\sqrt{\text{mean}} \tag{9.6}$$

From the keying sequence in Table A32 the warning limits are ± 7.59 and the action limits are ± 7.94. Thus, until cars have more than twenty-three faults there is no evidence of a significant drop in quality and conversely, until cars have fewer than seven faults, there has been no significant improvement.

5.2   *Problem 10 – Customer complaints*

A public utility gets on average twenty complaints a week about service calls. Investigation of the cause of the complaints leads to new procedures being adopted and at the end of the first week under the new system only nine complaints had been lodged. Is this below the lower control limits of the original system?

The answer is given in Appendix B.

6   Project variability

6.1   *Problem 11 – Probability of completing a project on time*

The critical path through a project is found to be the activities listed in Table 9.3. The optimistic, pessimistic and most likely times are as shown. What is the probability of completing the project in 52 weeks?

Table 9.3 Critical path durations

Activity	Optimistic time (weeks)	Pessimistic time (weeks)	Most likely time (weeks)
A	2	6	3
B	5	9	8
C	3	12	6
D	4	12	8
E	9	16	12
F	5	18	8

The expected time for each activity $= \dfrac{a + 4m + b}{6}$ (9.7)

where
$$a = \text{optimistic time}$$
$$m = \text{most likely time}$$
$$b = \text{pessimistic time}$$

The total project duration is the sum of the expected times along the critical path. In cumulating each expected time in the memory of the calculator, the $\div 6$ can be left until the end and applied to the cumulative total.

The keying sequence is shown in Table A33 and gives an expected time of 46.8 weeks.

The variance of each activity $= \left(\dfrac{b - a}{6}\right)^2$ (9.8)

From this equation, the total variance of the critical path can be found by summing the individual variances. The standard deviation of the critical path about its expected value of 46.8 weeks is then found from the square root of the net variance.

The keying sequence is given in Table A34. Again, because 6 is a common factor, $(b - a)^2$ is cumulated and finally divided by 36 (i.e. 6^2). The standard deviation is calculated to be 3.312. The probability of completing in 52 weeks or less can now be found from normal distribution tables where:

standardised deviation $= \dfrac{\text{target date} - \text{expected duration}}{\text{standard deviation}}$ (9.9)

The standardised deviation $= 1.57$ (*confirm*)

and from the normal distribution table in Appendix D, the probability of *not* exceeding this is 0.9418, i.e. 94%.

6.1   *Problem 12 – Revision of estimates*

After 18 weeks, activities *A*, *B* and *C* in Table 9.3 have been completed. What is the revised probability of completing in a further 34 weeks?

The answer is given in Appendix B.

10 Science and Engineering

1 Introduction

Problems in science and engineering involving the use of a calculator may be categorised as follows:

(a) Problems whose solutions result in the formulation and/or use of a simple formula. The problem itself may be easy or difficult to solve but the final calculation requires relatively few entries of numbers and operations into the calculator.

(b) Problems requiring the manipulation of more complicated formulae. These may result in more lengthy calculations and allow scope for determining the most efficient method to use.

(c) Experimental results which are used in calculations to form a table of intermediate or final results. The table may be used to find a mean or some other statistical measure. Alternatively, the table may be used to draw a graph (this is a method often used to smooth out experimental errors) resulting in further calculations, or the final result may be read off the graph.

The following examples of applications in science and engineering have been selected to illustrate the way a calculator may be used as a tool. Often a real application will require the use of a calculator at different stages in the interpretation of experimental results and other techniques such as the use of graphs are considered.

Many applications involve the tabulation of results and repetitive calculations, i.e. the same formula is used with different values. It is worth spending some time determining the most efficient method of calculation in these cases and it is helpful to draw up a list of instructions as shown in the key entry tables. Noting intermediate results can be useful when checking the calculation, or a further calculation may be carried out to check the final answer.

It has been assumed that the reader has some knowledge of science or engineering. A brief explanation of the background of each problem has been given to allow the study of a particular example from the calculator point of view. This should allow the reader to apply similar techniques to quite different problems.

2 The use of simple formulae

2.1 *Problem 1 – Production of magnesium by electrolysis*

Magnesium may be produced by the electrolysis of fused magnesium chloride. How long would a current of 2.57 A need to be passed to produce 1.5 g of magnesium?

2.1.1 *Method*

Two faradays of electricity are required to discharge one gramme-ion of magnesium, and since the atomic mass of magnesium is 24.32, 24.32 g of magnesium will be produced by this quantity of electricity.

The quantity of electricity used when a current of 1 ampere flows for 1 second is called a coulomb. 1 faraday is equal to 96 500 coulombs.

1.5 g of magnesium will be produced by

$$\frac{1.5 \times 2 \times 96\,500}{24.32} \text{ coulombs}$$

and this will take

$$\frac{1.5 \times 2 \times 96\,500}{24.32 \times 2.57} \text{ seconds}$$

The keying sequence is given in Table A35.

It will take 1 hour, 17 minutes and 12 seconds to produce 1.5 g of magnesium.

2.2   Problem 2 – *Current required for electrolysis*

By how much would the current need to be increased to produce the same quantity of magnesium in half an hour?

The answer is given in Appendix B.

2.3   Problem 3 – *Poisson's ratio for a short wire*

During the stretching of a wire, there is a lateral contraction of the wire. The ratio of the fractional lateral contraction to the fractional longitudinal extension is known as Poisson's ratio (σ).

This was found experimentally for a short wire by Searle's method in which the wire is set into oscillation horizontally and vertically. The periods of oscillation, T_1 and T_2, were measured in each case. Substitute these in the formula

$$\frac{T_2^2}{T_1^2} = 2(1 + \sigma) \qquad (10.1)$$

and calculate Poisson's ratio for the wire.

$$25T_1 = 36.1 \text{ seconds}$$
$$10T_2 = 21.5 \text{ seconds}$$

Transposition of the formula leads to,

$$\sigma = \frac{T_2^2}{2T_1^2} - 1 \qquad (10.2)$$

The keying sequence is given in Table A36. Poisson's ratio for a short wire was found to be 0.108.

2.4   Problem 4 – *Percentage composition by mass of* $Fe_7C_{18}N_{18}$

Calculate the percentage composition by mass of ferric ferrocyanide ($Fe_7C_{18}N_{18}$), given that atomic masses of Fe, C and N are 55.84, 12.01 and 14.008 respectively.

Formula mass $= (55.84 \times 7) + (12.01 \times 18) + (14.008 \times 18)$

$= 390.88 + 216.18 + 252.144$

$= 859.204$

Percent $Fe = \dfrac{390.88}{859.204} \times 100$

$C = \dfrac{216.18}{859.204} \times 100$

$N = \dfrac{252.144}{859.204} \times 100$

The keying sequence is given in Table A37.

Note: Add percentages to check calculation. Result should be 100% approximately.

The answers are as follows:

Percentage of iron $= 45.49\%$,

of carbon $= 25.16\%$,

and of nitrogen $= 29.34\%$

2.5 *Problem 5 – Empirical formula of a compound*

A compound has the following composition:

copper (atomic mass 65.57)   1.31 g

sulphur (atomic mass 32.06)   0.64 g

oxygen (atomic mass 16.00)   1.28 g

Find its empirical formula.

(a) Divide the mass of each element by its atomic mass to convert the mass to gramme-atoms. The results give the ratios in which the elements are combined.

(b) Divide the ratios by the smallest ratio to convert them into whole numbers.

The answer is given in Appendix B.

2.6   *Problem 6 – Heat of combustion of ethanol*

In an experiment, a quantity of water was heated by completely burning a known mass of ethanol. Calculate the heat of combustion in kJ/mol of ethanol from the following measurements:

$$\text{Mass of ethanol burnt} \qquad = \quad 0.36 \text{ g}$$
$$\text{Mass of water heated by ethanol} = 100 \text{ g}$$
$$\text{Rise in temperature of water} \quad = \quad 23.5° \text{ C}$$

Heat evolved = mass of water × its specific heat × its rise in temperature

$$= 100 \times 1 \times 23.5 \times 4.18 \text{ joules}$$

therefore when one gramme-mole of ethanol (46 g) is completely burned, the amount of heat evolved is

$$\frac{100 \times 23.5 \times 4.18 \times 46 \times 10^{-3}}{0.36} \text{ kJ/mol}$$

Heat of combustion = − amount of heat evolved.

(*a*) Perform a simple chain calculation, keying in the numbers in turn followed by the operation. Key in 46×10^{-3} as 46 EXP − 3.

(*b*) Divide by 0.36 and press = to obtain the result.

The answer is given in Appendix B.

3   More complicated formulae

3.1   *Problem 7 – Compression ratio of a car*

The compression ratio of a car is 7.3 to 1. If the initial temperature of the air and petrol vapour is 11.7°C, and the ratio of the specific heats of the mixture is 1.23, calculate the final temperature of the mixture. Assume the change is adiabatic.

For an adiabatic change to a gas

$$T . V^{\gamma-1} = \text{constant} \qquad\qquad (10.3)$$

where T is the absolute temperature (°K), V is the volume and γ is the ratio of the specific heats of the gas

$$\frac{V_1^{\gamma-1}}{V_2^{\gamma-1}} = \frac{T_2}{T_1}$$

$$T_2 = \frac{T_1 V_1^{\gamma-1}}{V_2^{\gamma-1}} = T_1\left(\frac{V_1}{V_2}\right)^{\gamma-1}$$

$$T_2 = (273 + 11.7) \times 7.3^{0.23}$$

The keying sequence is given in Table A38.

The final temperature of the mixture is 176.7°C.

3.2   Problem 8 – Adiabatic expansion of a gas

A gas expands adiabatically until its volume is doubled. Given that the initial temperature of the gas is 14.5°C and the ratio of the specific heats of the gas is 1.76, find the final temperature of the gas in °C.

$$T_2 = T_1(\tfrac{1}{2})^{0.76} = \frac{T_1}{2^{0.76}} \qquad (10.4)$$

The keying sequence is:

(a) log 2 × .76.
(b) antilog of result and place in memory.
(c) Enter 287.5 and divide by memory contents.
(d) Substract 273.

The answer is given in Appendix B.

3.3   Problem 9 – Molarity of nitric acid

A solution of 3.25 g of anhydrous sodium carbonate dissolved in 300 cm³ of water was used to neutralise 12 cm³ of nitric acid. What was the molarity of the nitric acid if it required 21.3 cm³ of the sodium carbonate solution to neutralise it?

$$\text{(atomic masses: sodium} = 22.997$$
$$\text{carbon} = 12.01$$
$$\text{oxygen} = 16.00)$$

Molarity of sodium carbonate solution =

$$\frac{\text{g in 1000 cm}^3 \text{ of solution}}{\text{gramme-formula mass of Na}_2\text{CO}_3} \qquad (10.5)$$

Equation for the reaction is

$$\text{Na}_2\text{CO}_3 + 2\text{HNO}_3 \rightarrow 2\text{NaCO}_3 + \text{CO}_2 + \text{H}_2\text{O} \qquad (10.6)$$

1 mole of sodium carbonate reacts with 2 moles of nitric acid.

1000 cm³ of nitric acid contains

$$\frac{1000}{12} \times 2\left(\frac{1000}{300} \times 3.25 \times \frac{1}{(2 \times 22.997) + 12.01 + (3 \times 16.00)}\right)$$
$$\times \frac{21.3}{1000}$$

The keying sequence is given in Table A39.

The molarity of the nitric acid was 0.363 M.

3.4 *Problem 10 – Molarity of sulphuric acid*

A solution of 5.75 g of potassium hydroxide dissolved in 250 cm³ of water is used to neutralise 15 cm³ of sulphuric acid. What is the molarity of the sulphuric acid if it required 36.4 cm³ of the potassium hydroxide solution to neutralise it?

$$\text{(atomic masses: potassium} = 39.096$$
$$\text{oxygen} = 16.00$$
$$\text{hydrogen} = 1.008)$$

Equation for the reaction is

$$2\text{KOH} + \text{H}_2\text{SO}_4 \rightarrow \text{K}_2\text{SO}_4 + \text{H}_2\text{O} \qquad (10.7)$$

The answer is given in Appendix B.

3.5   *Problem 11 – Moment of inertia of a wheel and axle*

The moment of inertia (I) of a wheel and axle may be found by rolling it down an inclined plane (see Figure 10.1).

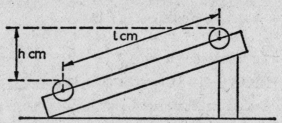

Figure 10.1   Moment of inertia of wheel and axle

$$mgh = \tfrac{1}{2}mv^2 + \tfrac{1}{2}I\left(\frac{v}{a}\right)^2 \tag{10.8}$$

where
$$m = \text{mass of wheel and axle}$$
$$a = \text{radius of axle}$$
$$v = \text{final velocity}$$

If t is the time of descent measured between two points a distance l apart, then the average velocity $= l/t$, and $v = 2l/t$

Therefore
$$I = ma^2\left(\frac{ght^2}{2l^2} - 1\right) \tag{10.9}$$

where g is the acceleration due to gravity.

The results of an experiment to measure I are as follows:

$m = 3.7\,\text{kg}$, diameter of axle $= 2.98\,\text{cm}$, $h = 1.2\,\text{cm}$, $t = 12.1\,\text{secs}$, $l = 70\,\text{cm}$.

Find I in kg/m^2 given $g = 9.81$ m/s^2

$$I = 3.7 \times \frac{2.98^2 \times 10^{-4}}{4}\left(\frac{9.81 \times 1.2 \times 10^{-2} \times 12.1^2}{2 \times 70 \times 70 \times 10^{-4}} - 1\right)$$

The keying sequence is given in Table A40.

The moment of inertia of the wheel and axle is 1.36×10^{-2} kg/m^2.

8.6 Problem 12 – Rydberg's constant for hydrogen

A numerical test of Bohr's theory of the hydrogen atom spectrum is provided by the Balmer series which allows the derivation of all the lines in the normal hydrogen atom spectrum by substituting integers for n in the following equation

$$\nu' = R_H \left(\frac{1}{2^2} - \frac{1}{n^2} \right) \qquad (10.10)$$

where R_H is the Rydberg constant for hydrogen and ν' is the reciprocal of the wavelength. The latter may be found using a spectrometer and plane diffraction grating.

In an experiment to determine the Rydberg constant, the angles of diffraction for the red, green and blue lines were

$$\theta_R = 22° \; 1'; \; \theta_G = 16° \; 8'; \; \theta_B = 14° \; 22'$$

The width of the diffraction grating (w) was 1.75×10^{-4} cm. Calculate Rydberg's constant for the red, green and blue lines and find the mean value, given that n has the values 3, 4 and 5 for the three lines respectively.

$$R_H = \left(\frac{1}{w \sin \theta} \right) \bigg/ \left(\frac{1}{4} - \frac{1}{n^2} \right) \qquad (10.11)$$

For the red line $R_H = \left(\dfrac{1}{1.75 \times 10^{-4} \times \sin 22° \; 1'} \right) \bigg/ \dfrac{5}{36}$

(a) Convert 22° 1' to degrees and find sine.
(b) Find reciprocal.
(c) Divide by 5 and multiply by 36.
(d) Repeat for green and blue lines.
(e) Find mean by adding the three values and dividing by 3.

The keying sequence is given in Table A41.

Rydberg's constant for the red line is 1.098×10^{-5} cm^{-1}.

For the remaining results see Appendix B.

3.7 Problem 13 – Refractive index of a glass prism

In an experiment to find the refractive index of a glass prism using a spectrometer, the following readings were taken:

Angle of minimum deviation $D = 224° 40' - 173° 13'$

Angle of prism A $2A = 245° 55' - 126° 9' = 119° 46'$

$$A = 59° 53'$$

Calculate the refractive index of the glass prism.

$$\text{Refractive index} = \frac{\sin \tfrac{1}{2}(A + D)}{\sin \dfrac{A}{2}} \qquad (10.12)$$

The keying sequence is given in Table A42.

The refractive index of the glass prism was 1.65.

4   Creating and using tables

4.1   Problem 14 – Finding the diameter of grains of lycopodium powder

In an experiment to find the average diameter of grains of lycopodium powder by the corona method, the radii of the first and second dark rings were measured for different values of l (see Figure 10.2). A is a hole in the screen through which light passed from a monochromatic (sodium) source.

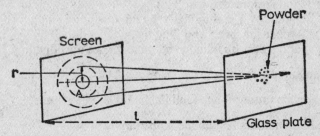

Figure 10.2   Diameter of grains of lycopodium powder

$$\theta_1 = 1.22\frac{\lambda}{a} \qquad\qquad \theta_2 = 2.233\frac{\lambda}{a} \qquad \begin{matrix}(10.13)\\(10.14)\end{matrix}$$

where θ_1 is the angle subtended by a diameter of the first dark ring and θ_2 by the second at a particle of lycopodium powder of radius a, and λ is the wavelength of the light source.

$$\theta_1 = \frac{2r_1}{l_1} \qquad\qquad \theta_2 = \frac{2r_2}{l_2} \qquad \begin{matrix}(10.15)\\(10.16)\end{matrix}$$

where r_1 and r_2 are the radii of the first and second dark rings respectively.

Note: $\frac{\theta_1}{2}$ and $\frac{\theta_2}{2}$ are small angles, therefore $\frac{\theta_1}{2} \simeq \tan\frac{\theta_1}{2}$

and similarly for $\frac{\theta_2}{2}$ (see p. 73).

Table 10.1   Values of l, r_1 and r_2 in cm

Experiment	l	r_1	θ_1	r_2	θ_2
1	12.9	0.292		0.523	
2	14.5	0.3288		0.5856	
3	15.9	0.362		0.642	

Mean values

Complete Table 10.1 and calculate the mean value of a, given that $\lambda = 5.893 \times 10^{-5}$ cm for the sodium light used.

$$\theta_1 = \frac{1.22\lambda}{a} \quad\text{and}\quad \theta_2 = \frac{2.233\lambda}{a}$$

Therefore
$$a_1 = \frac{1.22 \times 5.893 \times 10^{-5}}{\theta_1} \text{ cm} \qquad (10.17)$$

and

$$a_2 = \frac{2.233 \times 5.893 \times 10^{-5}}{\theta_2} \text{ cm} \qquad (10.18)$$

$$a = \frac{(a_1 + a_2)}{2}$$

(a) Calculate the first value of $\theta_1 = 2 \times r_1 \div l$ and store in memory. Calculate the other two values of θ_1 and add to memory. Recall the memory contents and divide by 3 to find the mean. Place the mean value in the memory.

(b) Calculate the numerator of (10.17). Divide by the contents of the memory to give a_1.

(c) Calculate θ_2 and the second value of a, and store in memory.

(d) Enter the first value and add to memory.

(e) Recall the memory contents and divide by 2 to give the mean value of a.

The keying sequence is given in Table A43.

The average diameter of the grains of lycopodium powder was 1.61×10^{-3} cm. The intermediate answers are given in Appendix B.

4.2   Problem 15 – Mutual inductance of a coil

Figure 10.3 shows a circuit set up to measure the mutual inductance (M) of a coil (primary L_1, secondary L_2). The resistance R_1 was kept at a particular value, and the variable resistance R was adjusted (to a value R_2) until no charge flowed through the

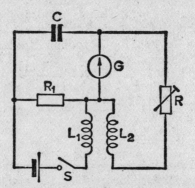

Figure 10.3   Circuit to measure mutual inductance of a coil

ballistic galvanometer G, on making or breaking contact with the switch S. The capacity of the condenser (C) was set to .6 μF.

The following formula applies

$$M = CR_1(R_2 + R_3) \tag{10.19}$$

where R_3 is the resistance of L_2.

Table 10.2 Values of R_2 and R_1 in ohms

R_2	R_1	$\dfrac{1}{R_1} \times 10^{-3}$
119	700	
182	600	
265	500	
389	400	
601	300	
1025	200	

Complete Table 10.2 (see Appendix B for values) and plot R_2 against $\dfrac{1}{R_1}$.

Find M in millihenries, and R_3 in ohms.
If the slope of the graph (Figure 10.4) is ϕ then, from (10.19),

$$\tan \phi = \frac{M}{C} = \frac{510}{2 \times 10^{-3}}$$

The intercept on the R_2 axis gives R_3.

$$M = \frac{510 \times .6 \times 10^{-6}}{2 \times 10^{-3}} \times 10^3 \text{ millihenries}$$

$$= 510 \times .3$$

$$= 153 \text{ mH}$$

Note: The calculator is useful in finding the reciprocals of R_1 to complete the table prior to drawing the graph. However, in this particular example, the results may be found simply from the graph.

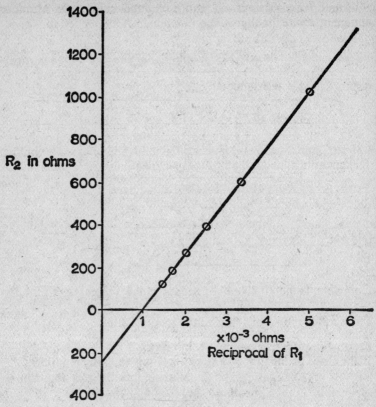

Figure 10.4   Graph of R_2 against $\frac{1}{R_1}$

4.3   *Problem 16 – Richardson's law*

In an experiment to test Richardson's law of thermionic emission, measurements were made of the anode current I_a and temperature T as shown in Table 10.3. I_a was measured at a particular value of V_a where saturation occurs in every case, so that Richardson's law

$$I_a = AT^2 e^{-b/T} \tag{10.20}$$

should apply.

A and *b* are constants. I_a is in milliamps and *T* is the absolute temperature of the filament.

Table 10.3 Values of I_a in milliamps and T in °K

I_a	11.2	19.1	34.2	56.7
T	2520	2585	2650	2715

Use the measurements given in Table 10.3 to determine whether Richardson's law holds in the experiment, and determine *b*.

From Richardson's equation (10.20) we have

$$\frac{b}{T} = \log_e A + \log_e \frac{T^2}{I_a}$$

or

$$\frac{1}{T} = \frac{1}{b}\log_e A + \frac{1}{b}\log_e \frac{T^2}{I_a}$$

Plot $\frac{1}{T}$ against $\log_e \frac{T^2}{I_a}$ as shown in Figure 10.5. A straight line confirms the form of Richardson's equation.

The slope of the graph $= \frac{1}{b}$.

Find the thermionic work function of the tungsten filament $= \frac{k \cdot b \times 10^{-8}}{e}$ electron volts,

where *k* is Boltzmann's constant $= 1.37 \times 10^{-16}$ and *e* is the electronic charge $= 1.59 \times 10^{-20}$.

(a) Enter first value of *T*, add to memory (ensure that the memory has been cleared first).
(b) Find the reciprocal of *T* and note in table.
(c) Recall memory and square.
(d) Divide by I_a.
(e) Find the log of $\frac{T^2}{I_a}$ and note in table.
(f) Clear memory and repeat for other values of *T* and I_a.

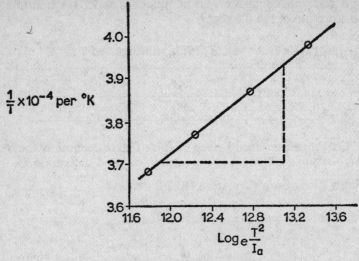

Figure 10.5   Graph of $\frac{1}{T}$ against $\log_e \frac{T^2}{I_a}$

(g) Plot graph and find slope. Find reciprocal to give b.
(h) Multiply by the given value of k and divide by the given value of e to find the thermionic work function.

The keying sequence is given in Table A44.

The results are given in Appendix B.

Appendix

Appendix A

Table A1   Checking bank statement

Keying	Display	Comments
6.45		
−		
20		no need to enter zero pence
+	−13.55	
50		
−	36.45	
20		
−	16.45	
.7		no need to enter final zero
+	15.75	
100		
−	115.75	
10		
−	105.75	
23.58		
=	82.17	final balance

Table A2   Checking telephone bill

Keying	Display	Comments
130		130 units
×		
.03		at 3p
+	3.9	cost of units
8.65		fixed charge
×	12.55	total less VAT
1.08		multiplier due to 8% VAT
=	13.55	total bill

Table A3 Kilometres per litre

Keying	Display	Comments
90		
—		
7		
÷	83.	purchase price
300		
×	0.27666	reciprocal of no. of 5 litres
157		
÷	43.436	km/5 litres
5		
=	8.6873	km/l

Table A4 Calculating the number of floor tiles for a given area

Keying	Display	Comments
340		length in cm
÷		
25		size of tile
=	13.6	14 tile lengths
260		width in cm
÷		
25		size of tile
=	10.4	11 tile lengths
11		whole number of tiles
×		
14		whole number of tiles
=	154.	number of tiles = 154

Table A5 Distance between landmarks

Keying	Display	Comments
47		
÷	47.	
60		
+	0.78333	
15		
=	15.783	15° 47′
M+		
cos	0.96229	cos 15° 47′
×		
428.6		
=	412.44	x = 412.44 metres
MR	15.783	
sin	0.27200	sin 15° 47′
×		
428.6		
=	116.57	h = 116.57 metres

Table A6 Stability of a lorry

Keying	Display	Comments
2.15		distance between wheels
÷		
2		
÷	1.075	
1.62		height of centre of gravity above level ground
=	0.66358	tan $\widehat{CAB}$
tan⁻¹	33.567	33°
−		
33		
×	0.56746	
60		
=	34.047	34′

Table A7 Area of triangle PQR

Keying	Display	Comments
237		
x^2	56169.	237^2
$\div$		
2		
$\times$	28084.5	
42		
sin	0.66913	sin 42°
$\times$	18792.1	
29		
sin	0.48480	sin 29°
$\div$	9110.6	
109		
sin	0.94551	sin 109° (alternatively, use sin 71°)
=	9635.5	area of $\triangle PQR$ = 9635.5 metres

Table A8 The radius of a circular track

Keying	Display	Comments
452		
x^2	204304.	452^2
+		
386		
x^2	148996.	386^2
−		
739		
x^2	546121.	739^2
$\div$	−192821.	
2		
$\div$	−96410.5	
452		
$\div$	−213.29	
386		
=	−0.55258	cos $P\widehat{Q}R$
$\cos^{-1}$	123.54	$P\widehat{Q}R$ = 123.54°
sin	0.83345	sin $P\widehat{Q}R$
$\dfrac{1}{x}$	1.1998	
$\times$		
739		PR
$\div$	886.66	
2		
=	443.33	radius = 443.3 metres

Table A9 Filling a swimming pool

Keying	Display	Comments
4.5		
x^2	20.25	4.5^2
÷		
4		2^2
×	5.0625	
π	3.1415926	
×	15.904	
1.75		
÷	27.832	
.12		
÷	231.93	
3.6		
=	64.427	64 hours
−		
64		
×	0.427	
60		convert decimal part to minutes
=	25.631	26 minutes

Table A10 Volume of a warehouse

Keying	Display	Comments
13		
÷		
18		
=	0.72222	
sin^{-1}	46.238	
×		
2		
=	92.476	$2 \sin^{-1}\left(\dfrac{13}{18}\right)$
M+		
×		
π	3.1415926	
÷	290.52	
180		
=	1.6140	$2 \sin^{-1}\left(\dfrac{13}{18}\right) \times \dfrac{\pi}{180}$
MEX	92.476	
sin	0.999066	$\sin\left(2 \sin^{-1}\left(\dfrac{13}{18}\right)\right)$
M+		$\left(2 \sin^{-1}\left(\dfrac{13}{18}\right) \times \dfrac{\pi}{180}\right) + \sin\left(2 \sin^{-1}\left(\dfrac{13}{18}\right)\right)$ in memory
2		
×		
π	3.1415926	
−	6.2831	
MR	2.6130	
=	3.6700	
×		
81		
÷	297.27	
2		
×	148.63	
52		
=	7729.2	volume = 7729 m^3

Table A11 Volume and mass of a wooden stand

Keying	Display	Comments
18		
x^2	324.	A_1
M+		
×		
13	13.	
x^2	169.	A_2
=	54756.	$A_1 \times A_2$
$\sqrt{x}$	234.	
+		
MR	324.	A_1
+	558.	
169		A_2
×	727.	
11		h
÷	7997.	
3		
=	2665.6	volume = 2666 cm³
EXP		
−		
6	2665.6 −06	
×	0.0026656	volume = 0.00267 m³
760		
=	2.0258	mass = 2.03 kg

Table A12 Calculating the number of locking pins per kilogram

Keying	Display		Comments
1.5			r_1
×			
2.5			r_2
+	3.75		
1.5			
x^2	2.25		r_1^2
+			
2.5			
x^2	6.25		r_2^2
×	12.25		
30			H
=	367.5		
M+			
3.5			
x^2	12.25		3.5^2
×			
3			
+	36.75		
5.5			
×	42.25		
5.5			
÷	232.37		
2			
=	116.18		
M+			
π	3.1415926		
×			
MR	483.68		
÷			
3			
×	506.51		
8900			density of metal kg/m^3
EXP			
−			
9	8900.	−09	
=	4.5079	−03	mass of each pin in kg
1/x	221.82		answer 222 pins/kg

Table A13 Calculation of coded standard deviation

Keying	Display	Comments
357		Σfx
x^2	127449.	$(\Sigma fx)^2$
÷		
185		n
=	688.91	
M+		
945		Σfx^2
−		
MR		
÷	256.08	
185		n
=	1.3842	variance
$\sqrt{x}$	1.1765	standard deviation = 1.18
×		
20		
=	23.5305	actual standard deviation = 23.53

Table A14 Area in tail of the Normal distribution

Keying	Display	Comments
2		x
×		
.23164		constants entered to five
+		significant figures
1		
=	1.46328	
1/x	0.68339	t
M+		store t
×		
1.3303		b_5
−	0.90905	− as b_4 is negative
1.8213		b_4
×	−0.91224	
MR		t
+	−0.62342	

Table A14—*continued*

Keying	Display	Comments
1.7814		b_3
×	1.1579	
MR		t
−	0.79135	− as b_2 is negative
.35656		b_2
×	0.43479	
MR		t
+	0.29713	
.31938		b_1
×	0.61651	
MR		t
=	0.42132	
MEX		store in place of t
2		x
x^2		
÷	4.	
2		
=	2.	
e^x	7.3890	$e^{x^2/2}$
1/x	0.13533	
×		
MR	0.42132	
=	0.05702	
MEX		store in place of 0.42132
2		
×		
π		
=	6.2831	
$\sqrt{x}$	2.5066	
1/x	0.39894	k
×		
MR	0.05702	
=	0.02274	Area in tail $= 0.0227$

Table A15 Exponential smoothing

Keying	Display	Comments
30		assumed 'old' value
×		
.8		(1 − 0.2)
=	24.	
M+		
30		1973 value
×		
.2		smoothing constant
+	6.	
MR	24.	
=	30.	smoothed average (for 1974)
×		
.8		
=	24.	
MEX		place revised value in memory
28		1974 value
×		
.2		
+	5.6	
MR	24.	
=	29.6	smoothed average (for 1975)
×		
.8		
=	23.68	
MEX		
34		1975 value
×		
.2		
+		
MR	23.68	
=	30.48	smoothed average (for 1976)
×		
.8		
=	24.384	
MEX		
32		1976 value
×		
.2		
+	6.4	
MR		
=	30.784	smoothed average (for 1977)

Table A15—*continued*

Keying	Display	Comments
×		
.8		
=	24.627	
MEX		
36		1977 value
×		
.2		
+	7.2	
MR		
=	31.827	smoothed average (for 1978) = 31.83

Table A16   Calculating a cash discount

Keying	Display	Comments
60		5 dozen
×		
2.15		price
=	129.	cost
M+		
24		2 dozen
×		
.4		price
=	9.6	cost
M+		
36		3 dozen
×		
3.1		price
=	111.6	cost
M+		
MR	250.2	total cost
×		
.05		5%
=	12.51	discount due
M—		
MR	237.69	Answer £237.69

Table A17 Price adjustment

Keying	Display	Comments
.7		wholesale price
M+	0.7	
×		
.6		% mark up
=	.42	
M+		
MR	1.12	selling price
×		
.25		% discount rate
=	0.28	discount
M−		
MR	0.84	sale price

Table A18 Price adjustment (short version)

Keying	Display	Comments
.7		wholesale price
×		
1.6		60% incease = 1.6 multiplier
×	1.12	selling price
.75		reduced to 75% of original
=	0.84	sale price

Table A19 Monthly repayments using y^x key

Keying	Display	Comments
1		
+		
.01		% interest, i, as a decimal
=	1.01	$(1 + i)$
y^x	0.00995	
240		no. of periods, 20 × 12 months
=	10.892	£10.89

154 *The Pocket Calculator*

Table A20 Monthly repayments using *ln* and e^x keys

Keying	Display	Comments
1		
+		
.01		i
=	1.01	$(1 + i)$
ln	0.00995	
×		
240		n
=	2.3880	
e^x	10.892	£10.89

Table A21 Monthly repayments using x^2 key

Keying	Display	Comments
1		
+		
.01		
=	1.01	$(1 + i)$
x^2	1.0201	
x^2	1.0406	
x^2	1.0828	
x^2	1.1725	bottom line and part of top
M+		store as bottom line
x^2	1.3749	continue as top line
x^2	1.8904	
x^2	3.5738	
x^2	12.772	top line
÷		
MR	1.1725	bottom line recalled
=	10.892	£10.89

Table A22 Monthly repayments; final part of calculation

Keying	Display	Comments
=	10.89	from previous methods
1/x	0.09182	
M+		
1		
−		
MR	0.09182	
=	0.90817	
1/x	1.1011	
×		
10000		Principal
×	11011.	
.01		*i*
=	110.11	£110.11

Table A23 Calculation of length of loan

Keying	Display	Comments
.125		interest rate p.a.
÷		
12		
=	0.01041	interest rate per month
+		
1		
=	1.01041	$(1 + i)$
M+		
−		
1		to return to interest rate again
×	0.01041	*i*
10000		*P*
÷	104.16	
200		
−	0.52083	
1		
=	−0.47916	$-(1 - Pi/R)$
x^2	0.22960⟩	to change sign without using
$\sqrt{x}$	0.47916⟩	memory
1/x	2.0869	
ln	0.73570	
MEX	1.01041	
ln	0.01036	
÷		
MR	0.73570	
=	0.01408	
1/x	70.995	70.9 months i.e. 5 years 11 months

Table A24 Calculating present worth factors

Keying	Display	Comments
1.15		$1 + i$
M+		
ln	0.13976	
×		
5		start at 5 years
=	0.69880	
e^x	2.01135	
1/x	0.497176	factor for 5th year = 0.49718
÷		
MR	1.15	
=	0.432327	factor for 6th year = 0.43233
÷		
MR	1.15	
=	0.375937	factor for 7th year = 0.37594
÷		
MR	1.15	
=	0.326901	factor for 8th year = 0.32690

Table A25 Rate of return

Keying	Display	Comments
6000		F
÷		
3000		P
=	2.	
ln	0.69314	log
÷		
7		7th root required
=	0.09902	
e^x	1.1040	anti-log
−		
1		
=	0.10408	× 100 = 10.4%

Table A26   Analysis of accounts

Keying	Display	Comments
2541		profit
M+		
÷		
6872		assets
=	0.36976	return on investment = 37%
MR	2541.	
÷		
12680		sales
=	0.20039	× 100 = % profit = 20%
12680		sales
÷		
4217		stock
=	3.0068	rate of stock turnover = 3.0

Table A27   Depreciation factor

Keying	Display	Comments
250		scrap value
÷		
6000		initial cost
=	0.04166	
ln	−3.1780	
÷		
10		no. of years
=	−0.31780	
e^x	0.7277444	10th root

Table A28   Break-even volume

Keying	Display	Comments
.5		variable cost A
−		
.2		variable cost B
=	0.3	
M+		
7500		fixed cost B
−		
3000		fixed cost A
=	4500.	
÷		
MR	0.3	
=	15000.	break-even volume

Table A29   Order quantity

Keying	Display	Comments
2		
×		
5		c
×	10.	
10000		d
÷	100000.	
.15		i
÷	666666.	
.18		p
=	3703703.	
$\sqrt{x}$	1924.5	order quantity = 1925

Table A30 Probability of stock-out

Keying	Display	Comments
260		h
M+		
+		
580		s
=	840.	$h + s$
MEX	260.	
÷		
MR	840.	
=	0.30952	Probability

Table A31 Machine utilisation

Keying	Display	Comments
1253		machine operating
÷		
2000		total observations
×	0.6265	
100		
=	62.65	% utilisation = P
M+		
100		
−		
MR	62.65	
×	37.35	$100 - P$
MR	62.65	P
÷	2339.9	
2000		N
=	1.1699	
√x	1.0816	
×		
2		
=	2.1633	% limits = 2.16

Table A32   Quality limits

Keying	Display	Comments
15		mean
$\sqrt{x}$	3.7829	
M+		$\sqrt{\text{mean}}$
×		
1.96		
=	7.5910	warning limits = 7.59
2.05		
×		
MR	3.8729	
=	7.9396	action limits = 7.94
MEX	3.8729	action limits into memory
15		
+		
MR	7.9396	
=	22.939	upper action limit = 23
−		
MR	7.9396	
=	15.	back to mean
−		
MR	7.9396	
=	7.0603	lower action limit = 7

Table A33 Expected duration of a project

Keying	Display	Comments
3		*m* for *A*
+		
8		*m* for *B*
+	11.	
6		*m* for *C*
+	17.	
8		
+	25.	
12		
+	37.	
8		
×	45.	
4		
+	180.	Σ4*m*
2		*a* for *A*
+	182.	
6		*b* for *A*
+	188.	
5		*a* for *B*
+	193.	
9		*b* for *B*
+	202.	
3		
+	205.	
12		
+	217.	
4		
+	221.	
12		
+	233.	
9		
+	242.	
16		
+	258.	
5		
+	263.	
18		
÷	281.	Σ(*a* + 4*m* + *b*)
6		
=	46.833	expected time = 46.8

Table A34   Standard deviation of the critical path

Keying	Display	Comments
6		b for A
—		
2		a for A
=	4.	$b - a$
x^2	16.	$(b - a)^2$
M+		
etc.		repeat for each activity
=	395.	$\Sigma(b - a)^2$
÷		
36		
=	10.972	variance
$\sqrt{x}$	3.3124	standard deviation = 3.312

Table A35   Production of magnesium by electrolysis

Keying	Display	Comments
1.5		
×	1.5	
2		
×	3.0	
96500		
÷	289500.	
24.32		
÷	11903.7	
2.57		
=	4631.8	answer in seconds
÷		
3600		convert to hours
=	1.2866	note 1 hour
—		
1		retain decimal part
×	0.28661	
60		convert to minutes
=	17.197	note 17 minutes
—		
17		retain decimal part
×	0.19703	
60		convert to seconds
=	11.822	note 12 seconds (rounded)

Table A36 Poisson's ratio for a short wire

Keying	Display	Comments
36.1		
÷		
25		
=	1.444	T_1
x^2	2.0851	$T_1{}^2$
×		
2		
=	4.1702	$2T_1{}^2$
M+		
2.15		T_2
x^2	4.6225	$T_2{}^2$
÷		
MR	4.1702	$2T_1{}^2$
−	1.1084	
1		
=	0.10844	$\sigma = 0.108$

Table A37 Percentage composition by mass of $Fe_7C_{18}N_{18}$

Keying	Display	Comments
55.84		
×		
7		
=	390.88	note for later keying-in
M+		
12.01		
×		
18		
=	216.18	note for later keying-in
M+		
14.008		
×		
18		
=	252.144	note for later keying-in
M+		
390.88		
÷		
MR	859.20	
=	0.45493	% Fe = 45.49
216.18		
÷		
MR	859.20	
=	0.25160	% C = 25.16
252.144		
÷		
MR	859.20	
=	0.29346	% N = 29.34
×		
100		
+	29.346	
45.49		
+	74.836	
25.16		
=	99.996	results checked

Table A38 Compression ratio of a car

Keying	Display	Comments
7.3		
ln	1.9878	$\log_e$ 7.3
×		
.23		
+	0.45721	
284.7		273 + 11.7 mentally
ln	5.6514	$\log_e$ 284.7
=	6.1086	
e^x	449.72	antilog$_e$ 6.1086
−		
273		convert to °C
=	176.72	answer 176.7°C

Table A39 Molarity of nitric acid

Keying	Display	Comments
2		
×		
22.997		
+	45.994	
12.01		
+	58.004	
48		3 × 16
=	106.00	
1/x	0.0094336	
×		
32.5		3.25 × 10 in numerator
÷	0.30659	
18		3 × 6
×	1.7032 −02	
21.3		
=	0.36280	molarity of nitric acid is 0.363 M

Table A40 Moment of inertia of a wheel and axle

Keying	Display		Comments
9.81			g
×			
1.2			
EXP			
−			
2	1.2	−02	h metres
×	0.11772		
12.1			t
x^2	146.41		t^2
÷	17.235		
2			
÷	8.6176		
70			l
÷	0.12310		
70			
EXP			
−			
4	70.	−04	70×10^{-4}
−	17.587		
1			
×	16.587		
3.7			m
×	61.372		
2.98			axle diameter in cm
×	182.88		
2.98			
EXP			
−			
4	2.98	−04	2.98×10^{-4}
÷	5.4501	−02	
4			
=	1.3625	−02	moment of inertia $= 1.36 \times 10^{-2}$ kg/m^2

Table A41 Rydberg's constant for hydrogen

Keying	Display		Comments
1			
÷			
60			
+	1.6666	−02	
22			
=	22.016		angle of diffraction for red line
sin	0.37487		sin 22° 1′
×			
1.75			
EXP			
−			
4	1.75	−04	
=	6.5602	−05	
1/x	15243.1		
÷			
5			
×	3048.6		
36			
=	109750.4		Rydberg constant = 1.098×10^{-5} cm^{-1}

Table A42   Refractive index of a glass prism

Keying	Display	Comments
53		
÷		
60		convert minutes to degrees
+	0.88333	
59		
=	59.883333	note for later entry
÷		
2		
=	29.941	$A/2$
sin	0.49911	$\sin A/2$
M+		
27		
÷		
60		
+	0.45	
51		
=	51.45	
+		
59.883333		
÷	111.33	
2		
=	55.666	$\frac{1}{2}(A + D)$
sin	0.82577	$\sin \frac{1}{2}(A + D)$
÷		
MR	0.49911	$\sin A/2$
=	1.6544	Answer 1.65

Table A43 Diameter of grains of lycopodium powder

Keying	Display		Comments
.292			r_1 1st value
×			
2			
÷	0.584		
12.9			l 1st value
=	4.5271	−02	θ_1 1st value
M+			
.3288			r_1 2nd value
×			
2			
÷	0.6576		
14.5			l 2nd value
=	4.5351	−02	θ_1 2nd value
M+			
.362			r_1 3rd value
×			
2			
÷	0.724		
15.9			l 3rd value
=	4.5534	−02	θ_1 3rd value
M+			
MR	0.13615		
÷			
3			
=	4.5385	−02	mean value = 0.0454
MEX			
1.22			
×			
5.893			
EXP			
−			
5	5.893	−05	5.893×10^{-5} cm (λ)
÷	7.1894	−05	
MR	4.5385	−02	
=	1.5840	−03	$a = 1.58 \times 10^{-3}$ cm

Table A44   Richardson's law

Keying	Display		Comments
2520			1st value of T
M+			
1/x	3.9682	-04	$1/T$
MR	2520.		
x^2	6350400.		T^2
$\div$			
11.2			1st value of I_a
$=$	567000.		T^2/I_a
ln	13.248		$\log_e \dfrac{T^2}{I_a}$

Appendix B

Answers to problems

Chapter 2

2. £49.01
4. 28% reduction
6. 0.23895 litres

Chapter 5

3. 24.8 cm
4. 38° 40′ or 141° 20′
7. 49.71 cm²
8. 0.045

Chapter 6

1. $l = w = \sqrt{\dfrac{v}{h}} = 4.0$ cm
4. 0.103 m³
5. 27.55 cm³

Chapter 7

2. assumed mean = 5
    scaling factor = 10
    Σfx = 240
    Σfx^2 = 570
    coded mean = 1.6
    coded S.D. = 1.114
    actual mean = 21.0
    actual S.D. = 11.14

5. r = 0.990
    m = 3.161
    c = −279
    residual = 3.809

8. 32.06

Chapter 8

2. £8.69

4. £102.50

8. .5787, .4823, .4019

10. £5484

11. £12 091

13. Return on investment = 40%
    % profit = 19%
    Rate of stock turnover = 3.4

15. .7952

Chapter 9

2. £1.00 price, using plastic process

4. 1491, say 1500

6. 4 (probability ⩽ 0.41)

8. Present accuracy = ±0.59%
   829 more readings

10. Yes (lower action limit was 10.8)

12. expected duration = 29.33
    standard deviation = 2.799
    probability = 0.9625

Chapter 10

2. 4.04 A

5. $CuSO_4$

6. −1255.16 KJ/mol

8. −103.23°C

10. 0.497 M

12. Rydberg constant for green and blue lines:
    1.0968×10^{-5} cm^{-1} and 1.0967×10^{-5} cm^{-1}
    Mean value = 1.0969×10^{-5} cm^{-1}

14. $\theta_2 = 0.0811, 0.0808, 0.0808$, mean value = 0.0809
    $a = 1.63 \times 10^{-3}$ cm
    Mean value of $a = 1.61 \times 10^{-3}$ cm

15. $\frac{1}{R_1} \times 10^{-3}$: 1.43, 1.67, 2.00, 2.50, 3.33, 5.00

16.

$\frac{1}{T} \times 10^{-4}$	3.97	3.87	3.77	3.68
$\log_e \frac{T^2}{I_a}$	13.248	12.765	12.232	11.775

$b = 1.2 \times 10^4/(3.935 - 3.705) = 5.217 \times 10^4$

Thermionic work function = 4.50 electron volts

Appendix C Constants and conversions

Constants

$$\pi = 3.1415926$$
$$e = 2.7182818$$
$$\log_e x = \log_e 10 \,.\, \log_{10} x = 2.302585 \log_{10}$$

Conversions to meteric

Given	Multiply by	To obtain
in	2.54	cm
ft	0.3048	metre
yd	0.9144	metre
mile	1.6093	km
areas: square the above factors		
volumes: cube the above factors		
oz	28.350	gramme
lb	0.45359	kg
ton (UK)	1.0160	tonne
ton (USA)	0.90718	tonne
fluid oz (UK)	28.412	ml
pint (UK)	0.56825	litre
gall (UK)	4.5460	litre
fluid oz (USA)	29.574	ml
pint (USA)	0.47318	litres
gall (USA)	3.7854	litres

Appendix D Normal distribution table

The quantity given is the probability that Z would be exceeded for a measurement drawn from a standardised normal distribution

Second Decimal Place of Z

Z	0	1	2	3	4	5	6	7	8	9
0.0	.5000	.4960	.4920	.4880	.4840	.4801	.4761	.4721	.4681	.4641
0.1	.4602	.4562	.4522	.4483	.4443	.4404	.4364	.4325	.4286	.4247
0.2	.4207	.4168	.4129	.4090	.4052	.4013	.3974	.3936	.3897	.3859
0.3	.3821	.3783	.3745	.3707	.3669	.3632	.3594	.3557	.3520	.3483
0.4	.3446	.3409	.3372	.3336	.3300	.3264	.3228	.3192	.3156	.3121
0.5	.3085	.3050	.3015	.2981	.2946	.2912	.2877	.2843	.2810	.2776
0.6	.2743	.2709	.2676	.2643	.2611	.2578	.2546	.2514	.2483	.2451
0.7	.2420	.2389	.2358	.2327	.2297	.2266	.2236	.2206	.2177	.2148
0.8	.2119	.2090	.2061	.2033	.2005	.1977	.1949	.1922	.1894	.1867
0.9	.1841	.1814	.1788	.1762	.1736	.1711	.1685	.1660	.1635	.1611
1.0	.1587	.1562	.1539	.1515	.1492	.1469	.1446	.1423	.1401	.1379
1.1	.1357	.1335	.1314	.1292	.1271	.1251	.1230	.1210	.1190	.1170
1.2	.1151	.1131	.1112	.1093	.1075	.1056	.1038	.1020	.1003	.0985
1.3	.0968	.0951	.0934	.0918	.0901	.0885	.0869	.0853	.0838	.0823
1.4	.0809	.0793	.0778	.0764	.0749	.0735	.0722	.0708	.0694	.0681
1.5	.0668	.0655	.0643	.0630	.0618	.0606	.0594	.0582	.0571	.0559

1.6	.0548	.0537	.0526	.0516	.0505	.0495	.0485	.0475	.0465	.0455
1.7	.0446	.0436	.0427	.0418	.0409	.0401	.0392	.0384	.0375	.0367
1.8	.0359	.0352	.0344	.0336	.0329	.0322	.0314	.0307	.0301	.0294
1.9	.0287	.0281	.0274	.0268	.0262	.0256	.0250	.0244	.0238	.0233
2.0	.0227	.0222	.0217	.0212	.0207	.0202	.0197	.0192	.0188	.0183
2.1	.0179	.0174	.0170	.0166	.0162	.0158	.0154	.0150	.0146	.0143
2.2	.0139	.0136	.0132	.0129	.0126	.0122	.0119	.0116	.0113	.0110
2.3	.0107	.0104	.0102	.00990	.00964	.00939	.00914	.00889	.00866	.00842
2.4	.00820	.00798	.00776	.00755	.00734	.00714	.00695	.00676	.00657	.00639
2.5	.00621	.00604	.00587	.00570	.00554	.00539	.00523	.00508	.00494	.00480
2.6	.00466	.00453	.00440	.00427	.00415	.00403	.00391	.00379	.00368	.00357
2.7	.00347	.00336	.00326	.00317	.00307	.00298	.00289	.00280	.00272	.00264
2.8	.00256	.00248	.00240	.00233	.00226	.00219	.00212	.00205	.00199	.00193
2.9	.00187	.00181	.00175	.00169	.00164	.00159	.00154	.00149	.00144	.00140
3.0	.00135	.00131	.00126	.00122	.00118	.00114	.00111	.00107	.00104	.00100
3.1	.00097	.00094	.00090	.00087	.00084	.00082	.00079	.00076	.00074	.00071
3.2	.00069	.00066	.00064	.00062	.00060	.00058	.00056	.00054	.00052	.00050
3.3	.00048	.00047	.00045	.00043	.00042	.00040	.00039	.00038	.00036	.00035
3.4	.00034	.00032	.00031	.00030	.00029	.00028	.00027	.00026	.00025	.00024
3.5	.00023	.00022	.00022	.00021	.00020	.00019	.00019	.00018	.00017	.00017

Source: From Table A, Appendix II of David Croft: *Applied Statistics for Management Studies,* published by Macdonald and Evans, London, 2nd edn, 1976. Reproduced by permission of the author and publishers.

Index

MATHEMATICS

L. C. Pascoe

This step-by-step introduction offers clear explanations and numerous worked examples that will guide the reader to an understanding of essential mathematical concepts and techniques.

Beginning with a brief historical outline of the development of mathematics, the book gently steers the reader through the basics of arithmetical processes, algebra and geometry. In keeping with the times, it then focuses on the electronic calculator both as a computational aid in financial calculations and other practical applications, and as a useful tool in progressing to more advanced mathematics. Thus problems involving percentages, profit and loss, and interest calculations are explored – and trigonometry introduced – using the functions available on most modern calculators. Throughout, exercises (with answers) are provided to test and reinforce the reader's understanding.

TEACH YOURSELF BOOKS